AF389004

Axisymmetry in Mechanical Engineering

Axisymmetry in Mechanical Engineering

Editor

Emanuel Willert

MDPI • Basel • Beijing • Wuhan • Barcelona • Belgrade • Manchester • Tokyo • Cluj • Tianjin

Editor
Emanuel Willert
Technische Universität Berlin
Germany

Editorial Office
MDPI
St. Alban-Anlage 66
4052 Basel, Switzerland

This is a reprint of articles from the Special Issue published online in the open access journal *Symmetry* (ISSN 2073-8994) (available at: https://www.mdpi.com/journal/symmetry/special_issues/Axisymmetry_Mechanical_Engineering).

For citation purposes, cite each article independently as indicated on the article page online and as indicated below:

LastName, A.A.; LastName, B.B.; LastName, C.C. Article Title. *Journal Name* **Year**, *Volume Number*, Page Range.

ISBN 978-3-0365-6489-0 (Hbk)
ISBN 978-3-0365-6490-6 (PDF)

Contents

About the Editor . . . . **vii**

Preface to "Axisymmetry in Mechanical Engineering" . . . . **ix**

Emanuel Willert
Guest Editorial: Special Issue "Axisymmetry in Mechanical Engineering"
Reprinted from: *Symmetry* **2023**, *15*, 174, doi:10.3390/sym15010174 . . . . **1**

Emanuel Willert
A Simple Semi-Analytic Contact Mechanical Model for Tangential and Torsional Fretting Wear
of Axisymmetric Contacts
Reprinted from: *Symmetry* **2021**, *13*, 1582, doi:10.3390/sym13091582 . . . . **3**

Chao-Ming Lin and Yun-Ju Chen
Coaxiality Optimization Analysis of Plastic Injection Molded Barrel of Bilateral Telecentric Lens
Reprinted from: *Symmetry* **2022**, *14*, 200, doi:10.3390/sym14020200 . . . . **15**

Valentin L. Popov
An Approximate Solution for the Contact Problem of Profiles Slightly Deviating from Axial
Symmetry
Reprinted from: *Symmetry* **2022**, *14*, 390, doi:10.3390/sym14020390 . . . . **41**

Ivan Argatov
A Comparison of General Solutions to the Non-Axisymmetric Frictionless Contact Problem
with a Circular Area of Contact: When the Symmetry Does Not Matter
Reprinted from: *Symmetry* **2022**, *14*, 1083, doi:10.3390/sym14061083 . . . . **53**

Alexander E. Filippov and Valentin L. Popov
Using Cylindrical and Spherical Symmetries in Numerical Simulations of Quasi-Infinite
Mechanical Systems
Reprinted from: *Symmetry* **2022**, *14*, 1557, doi:10.3390/sym14081557 . . . . **65**

Valentin L. Popov
Correction: Popov, V.L. An Approximate Solution for the Contact Problem of Profiles Slightly
Deviating from Axial Symmetry. *Symmetry* 2022, 14, 390
Reprinted from: *Symmetry* **2022**, *14*, 2108, doi:10.3390/sym14102108 . . . . **79**

About the Editor

Emanuel Willert

Dr.-Ing. Emanuel Willert studied Engineering Science and Mechanical Engineering at the Technische Universität (TU) Berlin and the Tomsk Polytechnic University. In 2019, he received his doctorate with honors from TU Berlin and the prize of the Dimitris N. Chorafas Foundation for his doctoral thesis on the fundamentals and applications of contact-impact problems. He is a co-author of the Handbook of Contact Mechanics and currently works as a post-doctoral researcher at the Institute of Applied Mechanics at TU Berlin. His main research interests are dynamic contact problems of inhomogeneous and inelastic materials, as well as the fundamentals of friction and wear.

Preface to "Axisymmetry in Mechanical Engineering"

The aim of the present Special Issue is to emphasize the phenomena, challenges and opportunities associated with symmetry in mechanical engineering. It is mainly addressed to researchers, scholars and practitioners in various fields of physics and engineering science. Several authors from different countries contributed to this Special Issue.

I, as a guest editor, am grateful to Ms. Flora Wang and the Symmetry production team for the management of the Special Issue and to the Symmetry editorial board, as well as to several reviewers for their scientific support.

Emanuel Willert
Editor

Editorial

Guest Editorial: Special Issue "Axisymmetry in Mechanical Engineering"

Emanuel Willert

Institute of Applied Mechanics, Technische Universität Berlin, Straße des 17. Juni 135, 10623 Berlin, Germany; e.willert@tu-berlin.de; Tel.: +49-30-314-21494

Citation: Willert, E. Guest Editorial: Special Issue "Axisymmetry in Mechanical Engineering". *Symmetry* **2023**, *15*, 174. https://doi.org/10.3390/sym15010174

Received: 3 January 2023
Accepted: 4 January 2023
Published: 6 January 2023

Axisymmetric (or almost axisymmetric) systems are ever-present in mechanical engineering. Axial symmetry enables free rotation, rolling and avoids field singularities at sharp corners. Spheres minimize the ratio of surface area to volume. Rotational symmetry (inherent to the system or enforced externally by the modelling) significantly simplifies the exact or numerical analysis of a model or, in some cases, only allows it in the first place. This list could go on; in fact, registering all the properties and applications of axisymmetry in technical or biotechnological systems would be quite an overwhelming task.

The topic of this Special Issue is thus very broad and, at the same time, very specific. Broad, because axisymmetry is ubiquitous in mechanical engineering, and specific, because it is a very particular mathematical and/or physical concept. This apparent disparity of thematic broadness and narrowness adds a special charm to the Special Issue, as the readers might discover similar specific methods and techniques in very different contexts, and apply them to yet again very different classes of problems—which has always been one of the main propellers of scientific discovery.

This Special Issue consists of five original research articles and one correction.

It opens with the original research paper "A Simple Semi-Analytic Contact Mechanical Model for Tangential and Torsional Fretting Wear of Axisymmetric Contacts" [1]. In this paper, the rotational symmetry of the problem allows for an asymptotic closed-form solution of the contact problem in each fretting oscillating cycle if the wear process is slow compared to the oscillation. In the case of wear laws, which explicitly relate the local wear intensities to contact stresses and relative surface displacements, this leads to a simple integro-differential equation for the wear contact problem. The model can be extended to incorporate the analysis of subsurface stress fields and associated initiation of fatigue cracks.

In the paper "Coaxiality Optimization Analysis of Plastic Injection Molded Barrel of Bilateral Telecentric Lens" [2] the authors study the optimization of process parameters for the injection moulding of cylindrical plastic lens barrels for bilateral telecentric lenses, based on the Taguchi method.

The original research article "An Approximate Solution for the Contact Problem of Profiles Slightly Deviating from Axial Symmetry" [3] provides an asymptotic closed-form solution for the normal contact problem of convex profiles that are almost axisymmetric. Thereby, the term "almost axisymmetric" also applies to profiles which one would not immediately associate with rotational symmetry, e.g., a pyramid with a square base. In this sense, the presented solution can be understood as a general (asymptotic) solution for arbitrary convex contact profiles.

Another important contact mechanical problem is studied in the research paper "A Comparison of General Solutions to the Non-Axisymmetric Frictionless Contact Problem with a Circular Area of Contact: When the Symmetry Does Not Matter" [4]. In the paper, the contact area is assumed to be fixed and circular, e.g., by using a cylindrical rigid punch for the indentation procedure, but the profile of the punch is arbitrary. Thus, one might say, that the contact problem for arbitrary convex profiles is "tackled from another side" through the lens of axial symmetry. The paper presents a deep comparative review of known closed-form and series solutions for the described contact problem and gives novel results for the pressure in the contact centre, as well as for the stress intensity factor at the contact boundary, which, among others, is important for the corresponding adhesive problem.

This Special Issue closes with the original research article "Using Cylindrical and Spherical Symmetries in Numerical Simulations of Quasi-Infinite Mechanical Systems" [5]. In this article, rotational symmetry is used as a tool to effectively remove simulation space boundaries—and, therefore, the corresponding artificial boundary effects—for the efficient numerical simulation of long-range interactions in general discrete periodic systems. Specifically, the mantle of a cylinder has no boundaries in the circumferential direction, while the surface of a sphere is boundary-free in all angular directions. As a very interesting example, the authors consider the tectonic dynamics of the earth's continents.

Finally, the correction [6] remediates misprints in the aforementioned original publication [3].

The authors who contributed to the Special Issue and the Guest Editor are hopeful that the published research here will facilitate a better understanding of the challenges and opportunities connected to axial (or other classes of) symmetry in different theoretical or applied branches of mechanical engineering.

Funding: This research was funded by the German Research Foundation under the project number PO 810/66-1.

Conflicts of Interest: The authors declare no conflict of interest.

References

1. Willert, E. A Simple Semi-Analytic Contact Mechanical Model for Tangential and Torsional Fretting Wear of Axisymmetric Contacts. *Symmetry* **2021**, *13*, 1582. [CrossRef]
2. Lin, C.-M.; Chen, Y.-J. Coaxiality Optimization Analysis of Plastic Injection Molded Barrel of Bilateral Telecentric Lens. *Symmetry* **2022**, *14*, 200. [CrossRef]
3. Popov, V.L. An Approximate Solution for the Contact Problem of Profiles Slightly Deviating from Axial Symmetry. *Symmetry* **2022**, *14*, 390. [CrossRef]
4. Argatov, I.I. A Comparison of General Solutions to the Non-Axisymmetric Frictionless Contact Problem with a Circular Area of Contact: When the Symmetry Does Not Matter. *Symmetry* **2022**, *14*, 1083. [CrossRef]
5. Filippov, A.E.; Popov, V.L. Using Cylindrical and Spherical Symmetries in Numerical Simulations of Quasi-Infinite Mechanical Systems. *Symmetry* **2022**, *14*, 1557. [CrossRef]
6. Popov, V.L.Correction: Popov, V.L. An Approximate Solution for the Contact Problem of Profiles Slightly Deviating from Axial Symmetry. *Symmetry* **2022**, *14*, 390. *Symmetry* **2022**, *14*, 2108. [CrossRef]

Article

A Simple Semi-Analytic Contact Mechanical Model for Tangential and Torsional Fretting Wear of Axisymmetric Contacts

Emanuel Willert

Institut für Mechanik, Technische Universität Berlin, 10623 Berlin, Germany; e.willert@tu-berlin.de;
Tel.: +49-30-314-21494

Abstract: Fretting wear of axisymmetric contacts is considered within the framework of the Hertz–Mindlin approximation and the Archard law for the linear wear. If the characteristic time scale for the wear is much larger than the duration of a single fretting oscillation, the profile change due to wear during one fretting cycle can be neglected for the contact problem as a zero-order approximation. This allows to give an exact contact solution during each fretting cycle, depending on the current worn profile, and thus for the explicit statement of an ordinary integro-differential equation system for the time-evolution of the fretting profile, which can be easily solved numerically. The proposed method gives the same results as a known, contact mechanically more rigorous simulation procedure that also operates within the framework of the Hertz–Mindlin approximation, but works significantly faster than the latter one. Tangential and torsional fretting wear are considered in detail. A comparison of the numerical prediction for the evolution of the worn profile in partial slip torsional fretting of a rubber ball on abrasive paper shows good agreement with experimental results from the literature.

Keywords: fretting wear; axis-symmetry; tangential contact; torsional contact; Abel transform

Citation: Willert, E. A Simple Semi-Analytic Contact Mechanical Model for Tangential and Torsional Fretting Wear of Axisymmetric Contacts. *Symmetry* **2021**, *13*, 1582. https://doi.org/10.3390/sym13091582

Academic Editor: Jan Awrejcewicz

Received: 11 August 2021
Accepted: 26 August 2021
Published: 27 August 2021

Publisher's Note: MDPI stays neutral with regard to jurisdictional claims in published maps and institutional affiliations.

1. Introduction

Many tribological contacts in technical or biological systems are subject to oscillations because they are part of a periodically operating machine or a periodically changing environment. The displacement amplitudes are often too small to cause global sliding, i.e., the contacts consist of changing areas of local sticking and slipping/sliding. This phenomenon, called "fretting", causes wear as well as fatigue and, although highly localized, is of enormous importance for the service life of various tribological systems, e.g., electrical conductors [1], turbine blades [2], or artificial joints [3].

Depending on the maximum extent of the slip area during the fretting oscillation, often different fretting regimes are distinguished [4], namely, the partial slip regime (for small slip areas), the gross sliding regime, and a mixed regime in between. While in different regimes, either wear or fatigue may be the dominating damage mechanism, both always are interacting or even competing phenomena [5]. Therefore, the theoretical solution procedure for a given fretting problem should consist of at least three components: a contact mechanical solution (depending on the material model, the frictional regime, and so on) as the basis for the determination of local stresses and displacements [6,7], a wear routine [8–10], as well as a crack initiation and propagation routine [11–13]. Doing this in a coupled and rigorous way (e.g., based on FE simulations [14]) is often very costly from a computational perspective. Here, semi-analytic (contact mechanical) approaches—which already have various applications in tribology [15]—can be a good "trade-off" between effort and prediction quality [16].

Therefore, in the present manuscript, a simple and efficient semi-analytic contact mechanical procedure for the fast simulation of fretting wear is proposed and discussed, which could also be extended for the analysis of fretting fatigue. The model operates within the framework of the Hertz–Mindlin approximation to general axisymmetric tangential contact problems, which will be stated in the manuscript, and which allows for an exact solution, if the contact problem only slowly changes with time.

The manuscript is organized as follows. In Section 2, first, the physical assumptions of the model are stated and the necessary classic solutions of axisymmetric contact mechanics are re-iterated. These solutions are formulated in terms of Abel-type integral transforms, the efficient numerical implementation of which is also briefly summarized. In Section 3, the semi-analytical procedure for the fast simulation of tangential and torsional fretting wear is presented and its predictions are compared to a contact mechanically more rigorous model and to experimental results from the literature. A discussion and conclusive remarks finish the manuscript.

2. Materials and Methods

2.1. Fundamental Assumptions

Firstly, we assume that all deformations are elastic. It follows from the fundamental solution by Boussinesq [17] and Cerruti [18] (see, e.g., the book of Johnson [19]) for the static relative surface displacements between two elastic half-spaces with the shear moduli G_1 and G_2 and the Poisson ratios v_1 and v_2, which act on each other with a point force at the origin of a common cartesian coordinate system $\{x,y,z\}$—z being the direction of the common normal of the half-spaces—that the normal and tangential contact problems are elastically decoupled, if the half-spaces are elastically similar, i.e., if

$$\frac{1-2v_1}{G_1} = \frac{1-2v_2}{G_2}.$$

$$(1)$$

For half-space contact mechanics to be valid, the contacting bodies must obey the restrictions of the half-space approximation, i.e., their macroscopic dimensions have to be much bigger than the largest characteristic contact size, and the surface gradients in the vicinity of the contact must be small. Moreover, surface roughness is neglected, and all time-dependent processes are analyzed in the quasi-static limit, i.e., elastic wave propagation is not considered. Moreover, friction in the contact shall obey a local Amontons–Coulomb law with a constant coefficient of friction μ, and for the uniaxial tangential contact problem with friction, the lateral displacements u_y resulting from a surface shear stress τ_{xz} are neglected. The entirety of the latter assumptions is what we will refer to as *Hertz–Mindlin approximation* [20,21] to the axisymmetric tangential contact problem with friction.

For the determination of the wear dynamics, a local Archard [22] wear law is used. Note that the Archard law, in connection with the friction law assumed by the contact mechanical component of the model, is equivalent to an energy-based wear model, which is commonly used in the literature on fretting wear [23]. However, from the point of view of the semi-analytical contact model given in Section 3, any local wear law according to which the profile change is an explicit function of the local contact pressure and the local slip displacement (or other known quantities) could be applied.

The severity of all these assumptions will be discussed later in the manuscript.

2.2. Elementary Axisymmetric Contact Solutions

Let the gap between the contacting bodies at first contact be an axisymmetric, monotonous, continuous function $f(r)$, with r being the polar radius in the contact plane. Then, the normal contact solution (with prescribed contact radius a) for the indentation depth d, normal force F, contact pressure distribution $p(r)$, and relative displacements $u_z(r)$ outside the contact area, first found by Schubert [24] and Galin [25], is given by [26]

$$d = g(a), \quad g(x) = \frac{\mathrm{d}}{\mathrm{d}x}\left[\int_0^x \frac{rf(r)}{\sqrt{x^2-r^2}}\mathrm{d}r\right], \quad F(a) = 2E^*\left[g(a)a - \int_0^a g(x)\mathrm{d}x\right],$$

$$p(r;a) = \frac{E^*}{\pi}\int_r^a \frac{g'(x)}{\sqrt{x^2-r^2}}\,\mathrm{d}x, \quad r \leq a, \quad u_z(r;a) = \frac{2}{\pi}\int_0^a \frac{g(a)-g(x)}{\sqrt{r^2-x^2}}\,\mathrm{d}x, \quad r > a,$$

$$(2)$$

with the effective Young's modulus

$$E^* := \left(\frac{1 - \nu_1}{2G_1} + \frac{1 - \nu_2}{2G_2} \right)^{-1}.$$

(3)

Usually, the prescribed quantities for the contact problem will rather be either the indentation depth (for displacement-controlled systems) or the normal force (for force-controlled systems), instead of the contact radius; however, the latter can easily be retained from the respective prescribed quantity by inverting the relevant equation in the above solution.

Now, consider the simple Cattaneo–Mindlin [21,27] loading history for the contacting bodies: they are first pressed against each other by a normal force F and subsequently loaded in the tangential direction by an increasing force F_x. As the tangential load increases, local slip propagates from the contact edge into the contact area. The radius of the remaining inner stick area at the end of the loading procedure shall be c. Ciavarella [28] and Jäger [29] were the first to recognize that the general axisymmetric Cattaneo–Mindlin problem (within the simplifications detailed in the previous subsection) is solved by a basis (indicated by the index "B") tangential stress distribution

$$|\tau_B(r; a, c)| = \mu \begin{cases} p(r; a) - p(r; c), & r \leq c, \\ p(r, a), & c < r \leq a, \end{cases}$$

(4)

i.e., the problem can be reduced to the frictionless normal contact problem. The slip displacement distribution in the slip area after the complete loading process is given by [26]

$$|\Delta u_{x,B}(r)| = \mu \frac{E^*}{G^*} [f(r) + u_z(r; c) - g(c)], \quad c < r \leq a,$$

(5)

with the effective shear modulus

$$G^* := \left(\frac{2 - \nu_1}{4G_1} + \frac{2 - \nu_2}{4G_2} \right)^{-1},$$

(6)

and the same notations as in Equation (2). The argument c in Equations (4) and (5) refers to the normal contact solution if the contact radius were c. The basis relations between c; the macroscopic relative tangential displacement, u_{x0}; and the tangential force, F_x, are [26]

$$u_{x0,B}(a, c) = \mu \frac{E^*}{G^*} [g(a) - g(c)], \quad F_{x,B}(a, c) = \mu [F(a) - F(c)].$$

(7)

The solution to the torsional contact problem is a little more elaborate numerically, because there is no equivalent to the Ciavarella–Jäger principle for tangential contact, i.e., the torsional contact problem with friction cannot be reduced to the frictionless normal contact. Nevertheless, conceptionally, the solution to the general axisymmetric torsional contact problem with friction poses no difficulties and has been found by Jäger [30]. The assumptions for that solution are comparable to the Hertz–Mindlin approximation, namely, the contact partners are assumed to be elastically similar, and the frictional interaction shall obey a local Amontons–Coulomb law.

For the elementary loading history (constant normal force, subsequently applied increasing torque; sometimes referred to as the "Lubkin [31] problem"), the solution goes as follows [32]: first, a function ϕ must be determined according to

$$\phi(a, x) = \frac{\mu}{2G} \int_x^a \frac{p(r; a)}{\sqrt{r^2 - x^2}} \, dr, \quad x \leq a, \quad G := \left(\frac{1}{G_1} + \frac{1}{G_2} \right)^{-1},$$

(8)

which gives the relation between the amplitude of the torsional oscillation, φ_A, and the stick radius:

$$\varphi_A = \phi(a, c).$$

(9)

After that, the slip displacement distribution in the slip area is given by

$$\Delta u(r; a, c) = \frac{4}{\pi r} \int_c^r \frac{x^2 [\varphi_A - \phi(a, x)]}{\sqrt{r^2 - x^2}} \, dx, \quad c < r \le a. \tag{10}$$

The torque as a function of the stick radius can be calculated easily, but will not be needed in later stages of the present manuscript.

2.3. Numerical Implementation of the Abel-Like Integral Transforms

The efficient numerical implementation of the integral transforms appearing in the previous subsection is straight forward, but shall be described briefly in the following.

Let us start with the determination of the equivalent profile $g(x)$ from the axisymmetric profile $f(r)$:

$$g(x) = \frac{dh}{dx}, \quad h(x) = \int_0^x \frac{r f(r)}{\sqrt{x^2 - r^2}} dr. \tag{11}$$

Following an idea by Benad [33], we integrate by parts to get rid of the singularity in the integrand:

$$h(x) = \int_0^x \frac{r f(r)}{\sqrt{x^2 - r^2}} dr = \int_0^x f'(r) \sqrt{x^2 - r^2} \, dr, \tag{12}$$

where the prime denotes the first derivative with respect to the given function argument. The last integral is evaluated on equidistant arrays for the spatial coordinates. Applying the trapezoidal rule, the integral is reduced to a simple matrix-vector multiplication. All derivatives are calculated by second-order finite differences.

Integration by parts for the pressure distribution $p(r)$ results in

$$p(r) = \frac{1}{\pi} \int_r^a \frac{g'(x)}{\sqrt{x^2 - r^2}} dx = \frac{1}{\pi} \left[\cosh^{-1}\left(\frac{a}{r}\right) g'(a) - \int_r^a \cosh^{-1}\left(\frac{x}{r}\right) g''(x) \, dx \right], \tag{13}$$

where $\cosh^{-1}$ denotes the inverse hyperbolic cosine. For the additional transformations in the torsional case, integration by parts gives

$$\phi(x) = -\frac{\mu}{2G} \int_x^a \cosh^{-1}\left(\frac{r}{x}\right) \frac{d}{dr} [p(r; a)] \, dr \tag{14}$$

and

$$\Delta u(r) = -\frac{2}{\pi} \int_c^r \left[x \sqrt{1 - \frac{x^2}{r^2}} + r \cosh^{-1}\left(\frac{x}{r}\right) \right] \frac{d}{dx} [\phi(a, x)] dx, \quad c < r \le a. \tag{15}$$

For brevity, the discretization procedure is omitted, as it works completely similar to as in the first case.

3. Results

In this section, the proposed semi-analytical procedure for the simulation of fretting wear is shown. First, the model for tangential fretting wear is detailed and its predictions are compared to a contact mechanically more rigorous numerical model in Sections 3.1 and 3.2. In Section 3.3., the main idea of the model is applied to a different fretting mode, torsional fretting, and the results of a numerical simulation for the evolution of the worn profile in a partial slip torsional fretting contact of a rubber ball on abrasive paper are compared to experiments from the literature in Section 3.4.

3.1. Model for Tangential Fretting Wear

The Cattaneo–Mindlin loading only corresponds to the first loading of the fretting contact. As is known, the solution to the tangential contact problem depends on the loading history. So, how will the contact solution look like during the fretting cycles? When the direction of the tangential motion is reversed, there is a moment of complete stick, because, according to the local Amontons law, the whole slip area is at the boundary state $|\tau| = \mu p$. Moreover, wear happens very slowly, i.e., on a much larger time scale than the fretting cycle duration. If there were no wear at all during two cycles, after reversing motion, a basis stress distribution can be linearly superposed [34]:

$$\tau(r) = \tau_B(r; a, c) - 2\tau_B(r; a, c^*), \tag{16}$$

to obtain the stress distribution with the current stick radius c^*. At the end of the half-cycle, the stress distribution is the same as at its beginning, with the sign reversed. Therefore, on the one hand, the relations between the amplitudes of the tangential force or displacement oscillation, F_A and u_A, and the minimum stick radius c of the fretting cycle are also given by Equation (7). On the other hand, the slip displacement during the half-cycle is

$$\Delta u_x(r) = 2\Delta u_{x,B}(r; a, c), \quad c < r \leq a. \tag{17}$$

During the second half-cycle, the same happens again. Hence, assuming a local Archard wear law, a zero-order approximation for the profile change during one cycle is given by

$$\Delta f(r, t) = \frac{4\mu E^*}{G^*} \frac{k_{\text{wear}}}{\sigma_0} p(r, t; a(t)) [f(r, t) + u_z(r; c) - g(c)], \tag{18}$$

where the time t is measured in cycle durations, k_{wear} is the dimensionless wear coefficient in the Archard law and σ_0 is the hardness of the worn material. "Zero-order approximation" in this context refers to the fact that the influence of the change of profile due to the wear in one cycle on the contact problem and the resulting change in the slip displacement is neglected.

Equations (17) and (18) are based on the supposition that there is a remaining stick area when reversing the tangential motion. At the gross slip transition, one obtains

$$\Delta u_x = 4\mu \frac{E^*}{G^*} f(r), \quad r \leq a, \quad \text{if } u_A = \mu \frac{E^*}{G^*} d := u_{x,c}. \tag{19}$$

If the sliding distances remain small compared with the contact radius (so that the system retains its approximate axial symmetry), the model can also be used in the gross slip regime. In the first fretting half-cycle, gross slip will start at a macroscopic relative tangential displacement

$$u_{x0} = u_A - 2u_{x,c}, \quad \text{if } u_A > u_{x,c}. \tag{20}$$

Hence, the local slip displacement distribution in the contact area after a full cycle is given by

$$\Delta u_x = 4\left[u_A - u_{x,c} + \mu \frac{E^*}{G^*} f(r)\right], \quad r \leq a, \quad \text{if } u_A > u_{x,c}. \tag{21}$$

Moreover, it is apparent from Equation (7) that, for both fully displacement-controlled (i.e., with a fixed indentation depth and amplitude of the tangential displacement oscillation) and fully force-controlled (i.e., with a fixed normal force and amplitude of the tangential force oscillation) tangential fretting, the minimum stick radius during one fretting cycle is constant over time, if there are no profile changes for $r < c$. That is self-consistent within the basic assumptions of the model (no wear without relative motion between contacting surface points), and c thus has the meaning of a permanent stick radius.

Hence, the proposed semi-analytical model works as follows: in every fretting cycle, the slip displacement distribution depending on the current profile and the current pressure distribution is calculated; based on the Archard law, the profile is updated and the new pressure distribution is calculated based on the normal contact solution. Finally, it should be noted that the second and third term in the slip displacement formula in Equation (5) only depend on the function g within the permanent stick area, which does not change over time.

Very common characteristics of fretting contacts are "fretting loops", i.e., hysteresis loops between the macroscopic relative tangential displacement and tangential force during the fretting process. The shape of the fretting loop and its time evolution are often used to characterize the fretting regime [35,36], because it is a simple and effective way to visualize the frictional energy dissipation in the slip area of the contact. The fretting loops can be calculated very easily in the model described above. From the basic contact solution in Equation (7) and the superposition in Equation (16), it is clear that the relative tangential displacement and the tangential force during a specific fretting cycle (without gross slip) with the current equivalent profile $g(x,t)$ are given by

$$
\begin{aligned}
F_x &= \pm F_{x,B}(a,c) \mp 2F_{x,B}(a,c^*) = \pm\mu[2F(c^*) - F(a) - F(c)], \quad c \le c^* \le a, \\
u_{x0} &= \pm u_{x0,B}(a,c) \mp 2u_{x0,B}(a,c^*) = \pm\mu\frac{E^*}{G^*}[2g(c^*) - g(a) - g(c)], \quad c \le c^* \le a,
\end{aligned}
\tag{22}
$$

where the different signs correspond to the first and second fretting half-cycles. For brevity, the time argument was omitted. In the gross slip regime, c equals zero and the tangential displacement during the partial slip stage of the oscillation must be modified according to

$$
u_{x0} = \pm\mu\frac{E^*}{G^*}[2g(c^*) - g(a)] \pm u_A \mp u_{x,c}, \quad 0 \le c^* \le a.
\tag{23}
$$

The force relation can be taken from Equation (22) and the relations in the sliding stage are trivial. Note that, because of the Ciavarella–Jäger principle, the fretting loop can be determined simply based on the normal contact solution for the current fretting cycle and the amplitude of the tangential oscillation.

3.2. Comparison with a Contact Mechanically More Rigorous Model

Dimaki et al. [37] proposed a fast numerical algorithm for fretting wear of axisymmetric contacts, where the slip displacement during one fretting cycle is calculated based on a separate (in the sense that it is almost independent of the wear simulation) numerical model within the framework of the method of dimensionality reduction (MDR) [38]; the fretting profile in their algorithm is also updated only once at the beginning of each fretting cycle and the MDR solution to the tangential contact problem operates within the Hertz–Mindlin approximation. Therefore, as the main (and practically only) difference between this numerical model and the semi-analytical approach proposed in the present manuscript is the calculation of the slip displacement, it makes sense to compare the predictions of both models regarding the time-evolution of the worn profile. In Figure 1, the results of both algorithms for the worn profiles are shown for an initially parabolic profile $f(r) = r^2/(2R)$, with some radius of curvature R, in normalized variables (note that, in normalized variables, the results only depend on the parameters given in the figure). Both algorithms were implemented in MATLAB® R2021a. The radial coordinate is normalized to the initial contact radius, $a_0 = (dR)^{1/2}$, and the worn profile for the fixed indentation depth d. The radius of the permanent stick area was set to $c = 0.5a_0$. The characteristic number of fretting cycles for the wear problem, N_0, results from the wear law [37]:

$$
N_0 = \frac{\pi a_0^2 d\sigma_0}{k_{\text{wear}} F_0 u_A},
\tag{24}
$$

with the initial normal force F_0 at the beginning of the fretting, and the tangential displacement oscillation amplitude u_A.

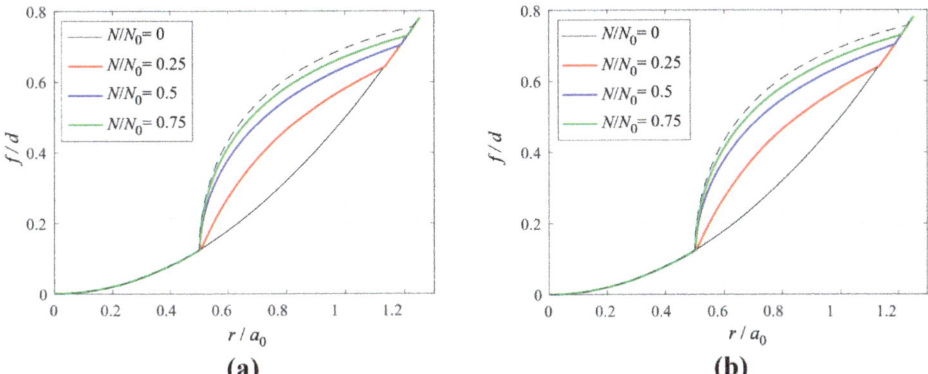

Figure 1. Time-evolution of the worn profile f, normalized for the fixed indentation depth d, as a function of the radial coordinate r, normalized for the initial contact radius a_0, for tangential fretting wear of an initially parabolic indenter and different numbers of normalized cycles, N/N_0. The dashed black line denotes the limiting no-wear profile [39]. The radius of the permanent stick area was set to $c = 0.5a_0$. (**a**) Results for the semi-analytic method proposed in the present manuscript; (**b**) results for the contact-mechanically more rigorous algorithm proposed by Dimaki et al. [37].

There are no visible differences between the prediction of the semi-analytical approach and the contact mechanically more rigorous model. Hence, it can be concluded that the zero-order approximation in Equation (18) gives a very good result (within the Hertz–Mindlin framework, of course) if the profiles are worn slowly compared with the fretting cycle duration. Nevertheless, if the implementation of the Abel-like transforms is done the same way, the semi-analytical model works faster by a factor of the order of one, because the number of numerical operations for one fretting cycle is significantly reduced.

3.3. Model for Torsional Fretting Wear

The main idea of the semi-analytical approach described in Section 3.1—that the slip displacement during one fretting cycle can be calculated analytically as a zero-order approximation—is also applicable to other fretting modes, for example, torsional fretting.

If, once again, we neglect the profile change on the time scale of a single fretting cycle, the torsional stress distribution between two reversal points of the fretting oscillation can be obtained by linearly superposing a stress distribution for the elementary loading in a similar way as in Equation (16) [40]. During one fretting cycle, this happens twice, so the total slip displacement during one fretting cycle is four times the expression given in Equation (10). The slip displacement once again determines the profile change according to the linear Archard wear law. Note that, for the torsional problem, $c = 0$ only exists as a limit, as the center point of the contact always remains sticking.

Hence, the semi-analytical algorithm for torsional fretting wear works the same as in the case of tangential fretting, except that now, four Abel-like transforms must be performed in every cycle. Interestingly, as the torsional contact problem cannot be reduced to the normal one, the minimal stick radius during one cycle is not constant while the profile is worn, but rather slightly increases. Nevertheless, the notation c shall still refer to the radius of the permanent stick region, i.e., the minimal stick radius during the first fretting cycle. The characteristic number of fretting cycles is given by

$$N_0 = \frac{\pi a_0 d \sigma_0}{k_{\text{wear}} F_0 \varphi_A},$$

$$(25)$$

with the amplitude of the torsional angular oscillation, φ_A.

3.4. Comparison with Experimental Results

In their analysis of the limiting no-wear profile for multi-mode fretting, Dmitriev et al. [41] conducted experiments on torsional fretting of a rubber half-sphere on a flat covered by abrasive paper. The radius of the sphere is (approximately) 19 mm (the paper wrongly says 14.5 mm, but from the respective figure, the correct value can be estimated), the indentation depth is 2 mm, and the radius of the permanent stick area is 3.25 mm [41]. The theoretical prediction will only depend on the number of normalized fretting cycles, N/N_0. In the experiments, the profiles were measured in consecutive steps of 2.160 cycles each. Based on the experiments, the value of N_0 was estimated as 14.400, which corresponds to a dimensional wear rate of approximately 7.5 mm^3/m (which is quite high). In Figure 2, the experimental results and the theoretical prediction are shown. They agree well with each other; in the experiments, the wear seems to be slower than predicted in the beginning (which makes sense, as the wear rate in the simulation is high); later, the wear rate in the experiments seems to increase and the convergence to the limiting profile is even faster than in the numerical calculation. Moreover, for the elastic ball in the experiment, there is a visible displacement of the lower point of the wearing area (corresponding to the edge of the permanent stick zone), while the position of the upper point (corresponding to the edge of the initial contact area) is virtually unchanged. On the other hand, the inverse relationship is observed in the calculation by the semi-analytical model.

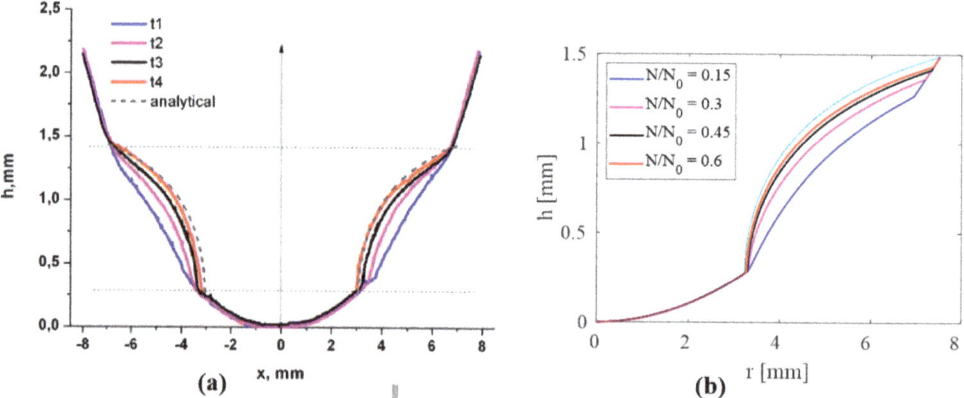

Figure 2. Time evolution of the worn profile of an initially spherical rubber indenter in torsional fretting on abrasive paper. (**a**) Experimental results by Dmitriev et al. [41] (figure by the authors of [41]); (**b**) theoretical prediction for consecutive numbers of normalized fretting cycles. The dashed line on the left and the thin solid line on the right correspond to the theoretical limiting no-wear profile [39].

4. Discussion

As stressed, the presented semi-analytical model of fretting wear is based on several simplifying assumptions that restrict its scope of applicability, at least in a quantitative sense. The most severe simplification is probably the local application of the linear Archard wear law. Moreover, neglecting surface roughness and non-elastic deformations can be unjustified simplifications in several circumstances.

Fouvry et al. [23] have pointed out that an elementary global-local, e.g., energy-based, description of the fretting wear process is applicable if the tribo-couple is dominated by non-adhesive wear mechanisms. This is confirmed by the above comparison of the present model proposal with the experimental results on torsional fretting wear of rubber on abrasive paper and in line with the findings of Varenberg et al. [42] that the role of oxide wear debris depends on the dominant fretting wear mechanism. However, in general, and especially in systems dominated by adhesive wear, an appropriate description of the third body in contact (i.e., the debris behavior) is vital for a comprehensive understanding of fretting wear. A deep overview on predictive wear models was given by Meng and Ludema [43]. For an up-to-date review on recent advances in the understanding of wear mechanisms and modelling, the reader is referred to the extensive work of Meng et al. [44]. However, quantitatively predictive wear equations are still extremely scarce and usually the main bottleneck for the predictive power of (even very elaborate) fretting wear simulations—which are often also based on a simple Archard- or energy-based wear law, despite their highly sophisticated material and contact models.

Another problematic characteristic of the model is that the macro-slip slope of the fretting hysteresis loop is zero, because it strictly is a Coulomb-based model, i.e., the friction coefficient is assumed to be constant (at least over one fretting cycle); however, in experimental fretting loops, the macro-slip slope is usually slightly positive (see, e.g., [45]).

Hu et al. [46] analyzed the effect of plastic deformations (due to the forming stress singularity at the boundary of the stick zone) on the fretting behavior and found that the main effect of plasticity is to allow the wear to penetrate the permanent stick region, and thus to never cease.

Moreover, the other assumptions of what in the present manuscript is referred to as the Hertz–Mindlin approximation can be a source of quantitative error. Generally, the tangential contact problem is not axisymmetric, and neither is, for example, the shape of the stick area. However, the error of the approximation is relatively small (at least compared with the uncertainty in the prediction of wear coefficients), as demonstrated by Munisamy et al. [47] by a self-consistent numerical solution of the Cattaneo–Mindlin problem. A deep and general analysis of tangential contact problems with elastic coupling can be found in the book of Barber [48].

On the other hand, several extensions of the proposed method are possible and planned for future work. Firstly, the Abel-like transforms can be written as explicit convolutions and, therefore, accelerated by the fast Fourier transform (FFT) [49]. Moreover, once the normal and tangential contact problems in the Hertz–Mindlin approximation are solved, the complete stress state inside the materials can be obtained via simple 1D-integrals [50], which allows for the fatigue analysis. Hence, the suggested semi-analytical model can be a part of a more global model of fretting in axisymmetric contacts. Finally, based on the tangential contact solution for power-law graded elastic materials by Heß and Popov [51], the proposed semi-analytical procedure can be generalized for power-law graded materials. This might be of high practical relevance, as this material class possibly offers a solution to the wear-fatigue dilemma in fretting [52].

5. Conclusions

A semi-analytic contact mechanical model for the fast simulation of fretting wear of axisymmetric contacts has been proposed. The main idea of the model is to neglect the change of the indenter profile during one cycle of the fretting oscillation, which—within the Hertz–Mindlin approximation—allows to analytically calculate the slip-length during one cycle in closed form, and hence gives a zero-order approximation for the time-evolution of the worn profile. This procedure gives the same results as contact mechanically rigorous simulations based on the Hertz–Mindlin approximation, but works significantly faster than those. A comparison of the numerical prediction for the evolution of the worn profile in partial slip torsional fretting of a rubber ball on abrasive paper shows good agreement with the experimental results from the literature.

Funding: This research was funded by the German Research Foundation under the project number PO 810/66-1.

Institutional Review Board Statement: Not applicable.

Informed Consent Statement: Not applicable.

Data Availability Statement: No additional data (other than stated in the manuscript) were produced or used for the preparation of the manuscript.

Conflicts of Interest: The author declares no conflict of interest. The funders had no role in the design of the study; in the collection, analyses, or interpretation of data; in the writing of the manuscript; or in the decision to publish the results.

References

1. Antler, M. Survey of contact fretting in electrical connectors. *Compon. Hybrids Manuf. Technol.* **1985**, *8*, 87–104. [CrossRef]
2. Rajasekaran, R.; Nowell, D. Fretting fatigue in dovetail blade roots: Experiment and analysis. *Tribol. Int.* **2006**, *39*, 1277–1285. [CrossRef]
3. Collier, J.P.; Mayor, M.B.; Jensen, R.E.; Surprenant, V.A.; Surprenant, H.P.; McNamar, J.L.; Belec, L. Mechanisms of failure of modular prostheses. *Clin. Orthop. Relat. Res.* **1992**, *285*, 129–139. [CrossRef]
4. Li, J.; Lu, Y.H. Effects of displacement amplitude on fretting wear behaviors and mechanism of Inconel 600 alloy. *Wear* **2013**, *304*, 223–230. [CrossRef]
5. Madge, J.J.; Leen, S.B.; Shipway, P.H. The critical role of fretting wear in the analysis of fretting fatigue. *Wear* **2007**, *263*, 542–551. [CrossRef]
6. Ciavarella, M.; Demelio, G. A review of analytical aspects of fretting fatigue, with extension to damage parameters, and application to dovetail joints. *Int. J. Solids Struct.* **2001**, *38*, 1791–1811. [CrossRef]
7. Goryacheva, I.G.; Rajeev, P.T.; Farris, T.N. Wear in Partial Slip Contact. *J. Tribol.* **2001**, *123*, 848–856. [CrossRef]
8. McColl, I.R.; Ding, J.; Leen, S.B. Finite element simulation and experimental validation of fretting wear. *Wear* **2004**, *256*, 1114–1127. [CrossRef]
9. Gallego, L.; Nélias, D. Modeling of Fretting Wear under Gross Slip and Partial Slip Conditions. *J. Tribol.* **2007**, *129*, 528–535. [CrossRef]
10. Pereira, K.; Yue, T.; Wahab, M.A. Multiscale analysis of the effect of roughness on fretting wear. *Tribol. Int.* **2017**, *110*, 222–231. [CrossRef]
11. Hills, D.; Nowell, D. *Mechanics of Fretting Fatigue*; Springer: Dordrecht, The Netherlands, 1994; 246p.
12. Fouvry, S.; Elleuch, K.; Simeon, G. Prediction of crack initiation under partial slip fretting conditions. *J. Strain Anal. Eng. Des.* **2002**, *37*, 549–564. [CrossRef]
13. Pereira, K.; Wahab, M.A. Fretting fatigue lifetime estimation using a cyclic cohesive zone model. *Tribol. Int.* **2020**, *141*, 105899. [CrossRef]
14. Llavori, I.; Zabala, A.; Urchegui, M.A.; Tato, W.; Gómez, X. A coupled crack initiation and propagation numerical procedure for combined fretting wear and fretting fatigue lifetime assessment. *Ther. Appl. Fract. Mech.* **2019**, *101*, 294–305. [CrossRef]
15. Aleshin, V.V. On Applications of Semi-Analytical Methods of Contact Mechanics. *Front. Mech. Eng.* **2020**, *6*, 30. [CrossRef]
16. Done, V.; Kesavan, D.; Chaise, T.; Nélias, D. Semi analytical fretting wear simulation including wear debris. *Tribol. Int.* **2017**, *109*, 1–9. [CrossRef]
17. Boussinesq, J. *Application des Potentiels a L'etude de L'Equilibre et du Mouvement des Solides Elastiques*; Imprimerie, L. Danel: Lille, France, 1885; 727p.
18. Cerruti, V. Ricerche intorno all'equilibrio de' corpi elastici isotropi. *Accad. Lincei Rend.* **1882**, *3*, 81–122.
19. Johnson, K.L. *Contact Mechanics*; Cambridge University Press: Cambridge, UK, 1985; 452p.
20. Hertz, H. Über die Berührung fester elastischer Körper. *J. die Reine Angew. Math.* **1882**, *92*, 156–171.
21. Mindlin, R.D. Compliance of Elastic Bodies in Contact. *J. Appl. Mech.* **1949**, *16*, 259–268. [CrossRef]
22. Archard, J.F. Contact and rubbing of flat surfaces. *J. Appl. Phys.* **1953**, *24*, 981–988. [CrossRef]
23. Fouvry, S.; Paulin, C.; Liskiewicz, T. Application of an energy wear approach to quantify fretting contact durability: Introduction of a wear energy capacity concept. *Tribol. Int.* **2007**, *40*, 1428–1440. [CrossRef]
24. Schubert, G. Zur Frage der Druckverteilung unter elastisch gelagerten Tragwerken. *Ing. Arch.* **1942**, *13*, 132–147. [CrossRef]
25. Galin, L.A. Three-Dimensional Contact Problems of the Theory of Elasticity for Punches with a Circular Planform. *Prikl. Mat. I Mekhanika* **1946**, *10*, 425–448.
26. Popov, V.L.; Heß, M.; Willert, E. *Handbook of Contact Mechanics: Exact Solutions of Axisymmetric Contact Problems*; Springer: Berlin/Heidelberg, Germany, 2019; 347p.
27. Cattaneo, C. Sul Contatto di due Corpore Elastici: Distribuzione degli sforzi. *Accad. Lincei Rend.* **1938**, *27*, 342–348, 434–436, 474–478.
28. Ciavarella, M. Tangential Loading of General Three-Dimensional Contacts. *J. Appl. Mech.* **1998**, *65*, 998–1003. [CrossRef]
29. Jäger, J. A New Principle in Contact Mechanics. *J. Tribol.* **1998**, *120*, 677–684. [CrossRef]
30. Jäger, J. Axi-symmetric bodies of equal material in contact under torsion or shift. *Arch. Appl. Mech.* **1995**, *65*, 478–487. [CrossRef]
31. Lubkin, J.L. The torsion of elastic spheres in contact. *J. Appl. Math. Mech.* **1951**, *73*, 183–187.
32. Willert, E.; Popov, V.L. Exact one-dimensional mapping of axially symmetric elastic contacts with superimposed normal and torsional loading. *J. Angew. Math. Mech.* **2017**, *97*, 173–182. [CrossRef]
33. Benad, J. Fast numerical implementation of the MDR transformations. *Facta Univers. Ser. Mech. Eng.* **2018**, *16*, 127–138. [CrossRef]
34. Jäger, J. Elastic contact of equal spheres under oblique forces. *Arch. Appl. Mech.* **1993**, *63*, 402–412. [CrossRef]
35. Varenberg, M.; Etsion, I.; Halperin, G. Slip Index: A New Unified Approach to Fretting. *Tribol. Lett.* **2004**, *17*, 569–573. [CrossRef]

36. Heredia, S.; Fouvry, S. Introduction of a new sliding regime criterion to quantify partial, mixed and gross slip fretting regimes: Correlation with wear and cracking processes. *Wear* **2010**, *269*, 515–524. [CrossRef]
37. Dimaki, A.V.; Dmitriev, A.I.; Chai, Y.S.; Popov, V.L. Rapid simulation procedure for fretting wear on the basis of the method of dimensionality reduction. *Int. J. Solids Struct.* **2014**, *51*, 4215–4220. [CrossRef]
38. Popov, V.L.; Heß, M. *Method of Dimensionality Reduction in Contact Mechanics and Friction*; Springer: Berlin/Heidelberg, Germany, 2015; 265p.
39. Popov, V.L. Analytic solution for the limiting shape of profiles due to fretting wear. *Sci. Rep.* **2014**, *4*, 3749. [CrossRef]
40. Jäger, J. Torsional impact of elastic spheres. *Arch. Appl. Mech.* **1994**, *64*, 235–248.
41. Dmitriev, A.I.; Voll, L.B.; Psakhie, S.G.; Popov, V.L. Universal limiting shape of worn profile under multiple-mode fretting conditions: Theory and experimental evidence. *Sci. Rep.* **2016**, *6*, 23231. [CrossRef]
42. Varenberg, M.; Halperin, G.; Etsion, I. Different aspects of the role of wear debris in fretting wear. *Wear* **2002**, *252*, 902–910. [CrossRef]
43. Meng, H.C.; Ludema, K.C. Wear models and predictive equations: Their form and content. *Wear* **1995**, *181–183*(Pt. 2), 443–457. [CrossRef]
44. Meng, Y.; Xu, J.; Jin, Z.; Prakash, B.; Hu, Y. A review of recent advances in tribology. *Friction* **2020**, *8*, 221–300. [CrossRef]
45. Yoon, Y.; Etsion, I.; Take, F.E. The evolution of fretting wear in a micro-spherical contact. *Wear* **2011**, *270*, 567–575. [CrossRef]
46. Hu, Z.; Lu, W.; Thouless, M.D.; Barber, J.R. Effect of plastic deformation on the evolution of wear and local stress fields in fretting. *Int. J. Solids Struct.* **2016**, *82*, 1–8. [CrossRef]
47. Munisamy, R.L.; Hills, D.A.; Nowell, D. Static Axisymmetric Hertzian Contacts Subject to Shearing Forces. *J. Appl. Mech.* **1994**, *61*, 278–283. [CrossRef]
48. Barber, J.R. *Contact Mechanics*; Springer: Basel, Switzerland, 2018; 585p.
49. Willert, E. FFT-based Implementation of the MDR transformations for homogeneous and power-law graded materials. *Facta Univers. Ser. Mech. Eng.* **2021**. [CrossRef]
50. Forsbach, F. Stress Tensor and Gradient of Hydrostatic Pressure in the Half-Space Beneath Axisymmetric Bodies in Normal and Tangential Contact. *Front. Mech. Eng.* **2020**, *6*, 39. [CrossRef]
51. Heß, M.; Popov, V.L. Method of Dimensionality Reduction in Contact Mechanics and Friction: A User's Handbook. II. Power-Law Graded Materials. *Facta Univers. Ser. Mech. Eng.* **2016**, *14*, 251–268. [CrossRef]
52. Willert, E.; Dmitriev, A.I.; Psakhie, S.G.; Popov, V.L. Effect of elastic grading on fretting wear. *Sci. Rep.* **2019**, *9*, 7791. [CrossRef]

 symmetry

Article

Coaxiality Optimization Analysis of Plastic Injection Molded Barrel of Bilateral Telecentric Lens

Chao-Ming Lin * and Yun-Ju Chen

Department of Mechanical and Energy Engineering, National Chiayi University, Chiayi 600355, Taiwan; s1053155@alumni.ncyu.edu.tw
* Correspondence: cmlin@mail.ncyu.edu.tw

Abstract: Plastic optical components are light in weight, easy to manufacture, and amenable to mass production. However, plastic injection molded parts are liable to shrinkage and warpage as a result of the pressure and temperature variations induced during the molding process. Consequently, controlling the process parameters in such a way as to minimize the geometric deformation of the molded part and improve the performance of the optical component as a result remains an important concern. The present study considered the problem of optimizing the injection molding parameters for the plastic lens barrel of a bilateral telecentric lens (BTL) containing four lens assemblies. The study commenced by using CODE V optical software to design the lens assemblies and determine their optimal positions within the barrel. Taguchi experiments based on Moldex3D simulations were then performed to determine the processing conditions (i.e., maximum injection pressure, maximum packing pressure, melt temperature, mold temperature, and cooling time) which minimize the coaxiality of the plastic barrel. Finally, CODE V and grayscale analyses were performed to confirm the optical performance of the optimized BTL. The Taguchi results show that the coaxiality of the plastic lens barrel is determined mainly by the maximum packing pressure and melt temperature. In addition, the CODE V and grayscale analysis results confirm that the optimized BTL yields a better modulus transfer function, spot diagram performance, and image quality than a BTL produced using the general injection molding parameters.

Keywords: injection molding; concentricity; coaxiality; optical barrel; bilateral telecentric lens; optimization

Citation: Lin, C.-M.; Chen, Y.-J. Coaxiality Optimization Analysis of Plastic Injection Molded Barrel of Bilateral Telecentric Lens. *Symmetry* **2022**, *14*, 200. https://doi.org/10.3390/sym14020200

Academic Editor: Emanuel Willert

Received: 18 December 2021
Accepted: 18 January 2022
Published: 20 January 2022

Publisher's Note: MDPI stays neutral with regard to jurisdictional claims in published maps and institutional affiliations.

1. Introduction

Plastic provides an ideal material for the fabrication of optoelectronic product accessories and supports due to its light weight, low cost, high mechanical strength, ease of processing, and good potential for mass production. However, the successful realization of plastic optical components involves many challenges, including optical design, integrated mechanism design, process parameter design, and product inspection and analysis. Plastic optical products are commonly fabricated using injection molding. However, while injection molding has significant advantages in terms of a high throughput, a low unit cost, and good repeatability, the quality of the molded components is extremely sensitive to the processing conditions, e.g., the injection pressure, packing pressure, melt temperature, mold temperature, packing time, and cooling time. Moreover, the processing parameters have both individual and interactive effects on the molding outcome. Consequently, determining the processing conditions which optimize the quality of the molded products (the dimensional tolerance, surface finish, mechanical strength, etc.) is extremely complex.

Optical components are subject to extremely tight manufacturing tolerances in order to achieve the necessary imaging quality. However, achieving these tolerances for plastic components realized through injection molding is extremely challenging, particularly in the case of geometric assemblies with high aspect ratios, or non-symmetrical geometries [1–4].

In many lens-based devices, the lens assemblies are housed within plastic bodies, and hence the problem of ensuring the imaging quality of the device involves not only guaranteeing the optical quality of the individual lenses, but also controlling the geometrical precision and form of the housing. For example, for bilateral telecentric lenses (BTLs), which are designed to provide a constant field of view (FOV) at all distances from the lens, roundness or concentricity errors of the lens barrel result in misalignments of the lens assemblies, which then seriously degrade the performance of the device. As a result, it is essential that the molding parameters are correctly controlled in such a way as to minimize the shrinkage, warpage, and residual stress of the barrel during the manufacturing process [5–7].

Many studies have demonstrated the feasibility of combining the Taguchi experimental method with mold flow simulations to determine the optimal injection molding processing parameters through a minimum number of experiential trials [8–11]. Furthermore, many software systems and algorithms have been applied to evaluate existing optical designs, optimize the performance of new imaging solutions, reduce manufacturing costs through process improvements, etc. [12–17]. Among the many optical design software packages currently available, CODE V (Synopsys, Inc., Mountain View, California, USA) is one of the most commonly used for the design, tolerancing, and performance evaluation of new optical configurations.

In the present study, a hybrid Taguchi/Moldex3D simulation approach was employed to optimize the injection molding processing parameters for the plastic lens barrel of the BTL device shown in Figure 1 consisting of four lenses and a protruding handle. Optimal analysis of plastic injection molding was based on the authors' previous research on the roundness and concentricity of the BTL [18]. The authors performed further optical analysis on the concentricity results to confirm the optimized results. CODE V optical software was first used to design the four lenses and to determine their optimal positions within the barrel. Taguchi/Moldex3D simulations were then performed to establish the injection molding conditions which minimize the coaxiality error of the lens barrel (i.e., the overall concentricity error of the four lenses) following the molding process. Finally, CODE V and grayscale analyses were performed to investigate the modulus transfer function (MTF), spot diagram characteristics, and imaging performance of the optimized BTL barrel.

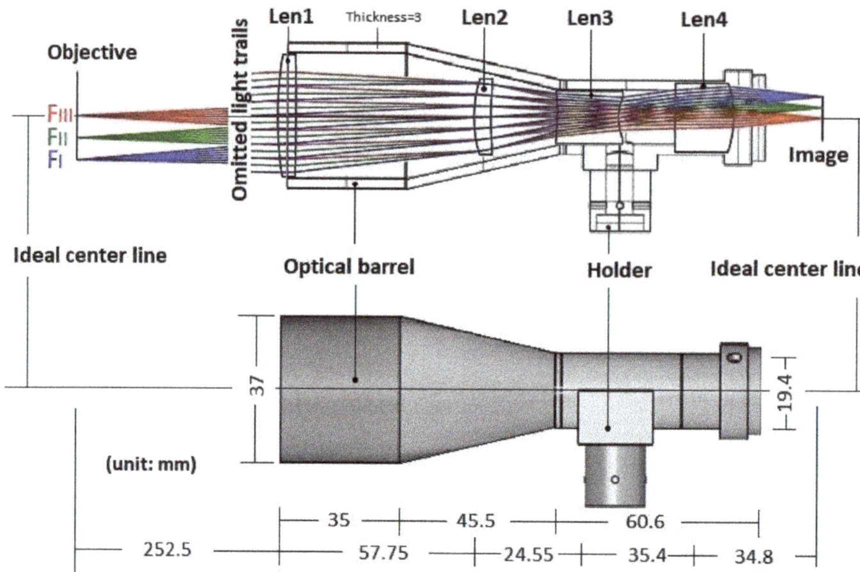

Figure 1. Geometry and optical design of the BTL (unit: mm).

2. Method and Procedures

Figure 2 presents a flow chart showing the main steps in the design, optimization, and analysis procedure performed in the present study. Having defined the optical components within the BTL (i.e., the four lenses), CODE V optical software was employed to establish the specification, location, and material of each lens and to determine the required geometrical dimensions of the optical tube. A model of the plastic optical tube was constructed using Rhinoceros software, and Moldex3D simulations were then performed in conjunction with the Taguchi experimental design method to determine the optimal injection molding conditions which minimize the average concentricity error of the optical tube at the planes corresponding to the ideal locations of the four lenses. Finally, the MTF and spot diagram of the BTL were investigated for three different light fields (F_I, F_{II}, and F_{III}) entering the objective lens at different distances from the optical axis (see Figure 1). The validity of the optimized design was confirmed by comparing the CODE V results for the modulus transfer function (MTF), spot diagram (SD), and 2D imaging simulation (IMS) of the proposed optical device with those of an ideal BTL device with no coaxiality or concentricity errors and a BLT device molded using the default processing conditions for the selected polymer material.

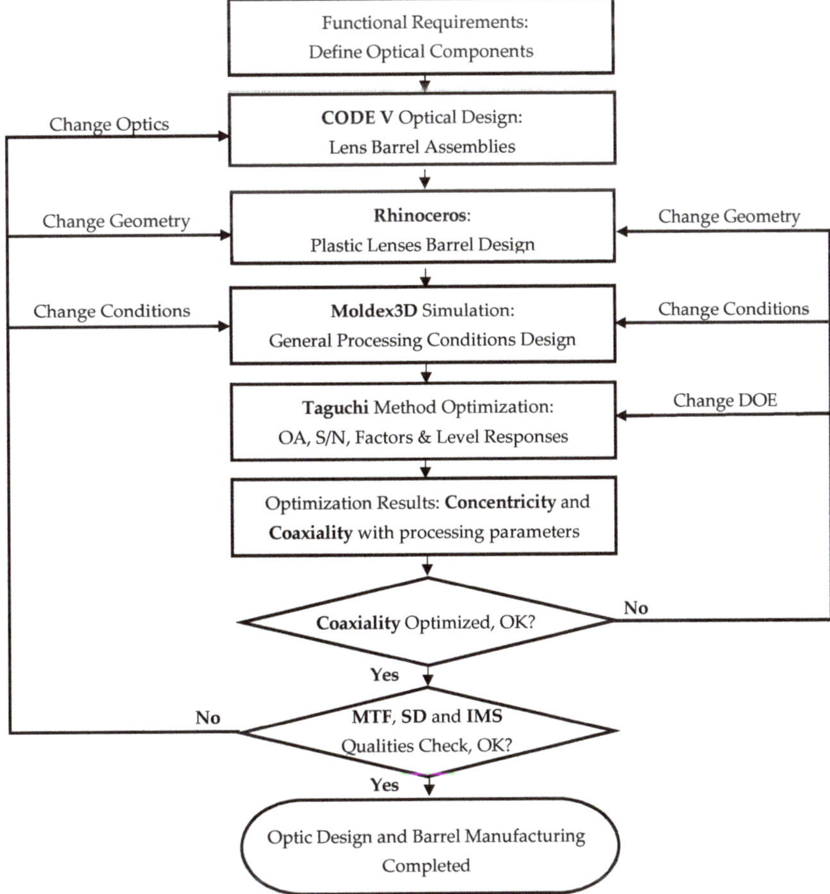

Figure 2. Flow chart showing the main steps in the optical design and injection molding optimization procedure for the BTL.

2.1. Materials and Geometry Model

In performing the analysis, it was assumed that the plastic barrel of the BTL was fabricated of PA66 (Polyamide 66 or Nylon 66) polymer (TECHNYL A 216, Solvay Engineering Plastics), with the material properties and recommended injection molding parameters shown in Table 1. Figure 3a shows the relationship between the viscosity of PA66 and the shear rate at different temperatures. Figure 3b shows the P-v-T (pressure/specific volume/temperature) properties of the PA66 material as a function of the shear rate and temperature, and this characteristic was used to calculate the degree of compression of the PA66 material in the packing stage. Table 2 shows the material, geometry dimensions, and optical properties of the four lenses within the BTL.

Table 1. Material properties and recommended processing conditions for PA66 (TECHNYL A 216, Solvay Engineering Plastics).

Material Properties	Values	Unit
Density	1140	Kg/m^3
Specific heat	2×10^{-11}	$J/Kg.\ ^\circ C$
Heat conduction coefficient	0.25	$J/s.m.\ ^\circ C$
Mold shrinkage	1.90	%
Tensile modulus	3000	MPa
Melt temperature	263	$^\circ C$
Coefficient of linear thermal expansion (23–85 $^\circ C$)	7×10^{-5}	$^\circ C$
Viscosity vs. shear rate under different temperatures	shown in Figure 3a	
P-v-T	shown in Figure 3b	
Processing Conditions in Injection Molding (Recommended)	**Values**	**Unit**
Maximum injection pressure	180–240	MPa
Maximum packaging pressure	180–240	MPa
Melt temperature	270–290	$^\circ C$
Mold temperature	60–100	$^\circ C$
Cooling time	11–17	s
Ejection temperature	190	$^\circ C$
Curing temperature	210	$^\circ C$

Table 2. Optical properties and geometrical dimensions of lenses in the BTL.

Lens No.	Glass/SCHOTT	Curvature Radius (mm)	Thickness (mm)	Refractive Index (-)	Abbe Value (-)	Partial Dispersion (-)	Dispersion (NL-N1) (-)
Len 1	NLAK33A	109.2083 −786.0770	5.0065	1.75393	52.271	0.30323	0.01442
Len 2	NLAK33B	38.8283 137.9732	5.0044	1.75500	52.300	0.30321	0.01444
Len 3	FK5HTI	−67.9914 15.3674	18.1944	1.48748	70.470	0.30977	0.00692
Len 4	LAFN7	427.9534 −33.0637	17.2165	1.74950	34.951	0.29414	0.02144

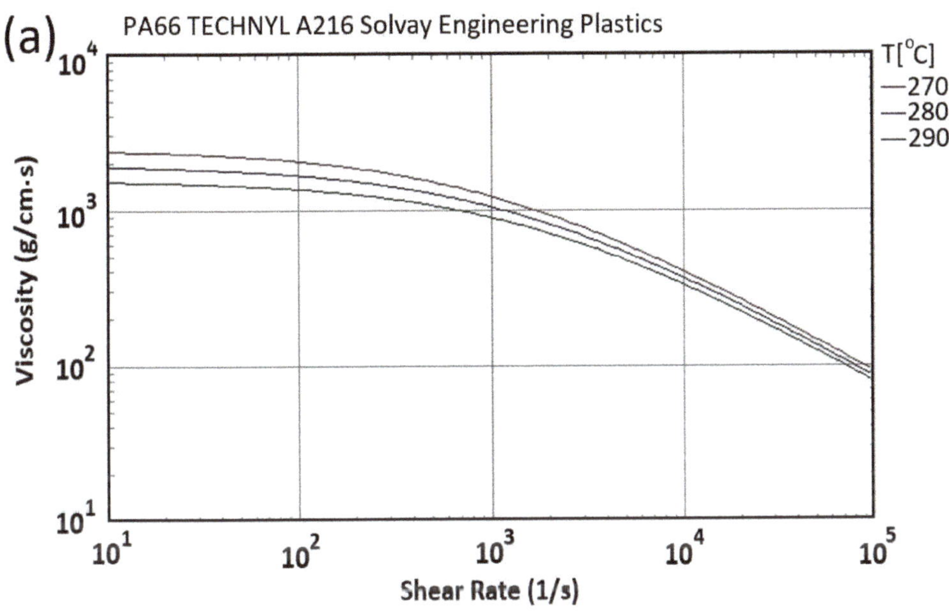

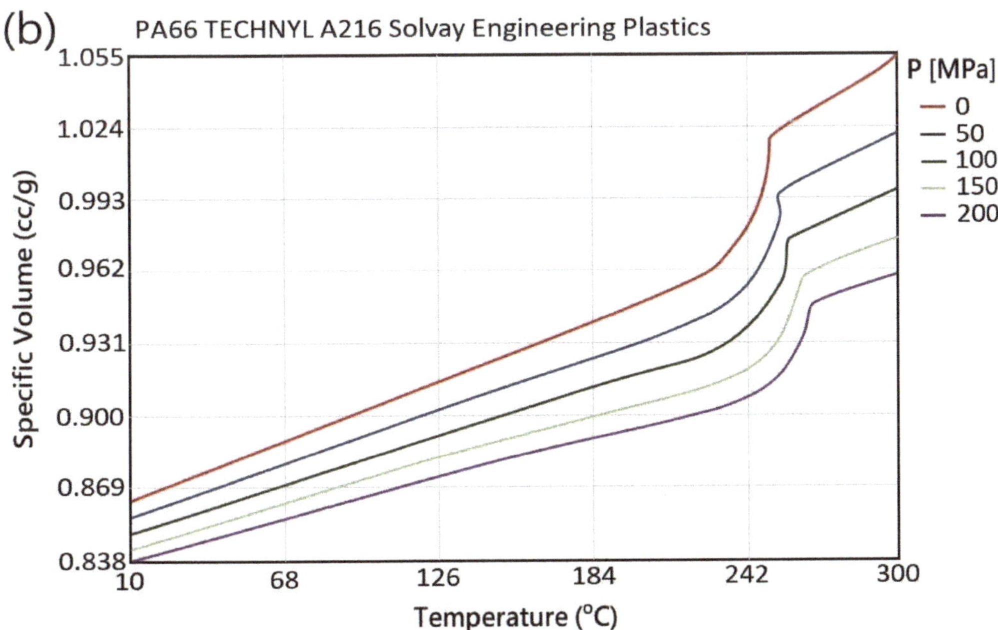

Figure 3. (**a**) Viscosity and (**b**) P-v-T properties of PA66 material (source: Moldex3D material library).

2.2. Mold Flow Analysis and Taguchi Experimental Method

A geometric model of the BTL barrel was constructed in Rhinoceros and exported to Moldex3D-Mesh to define the corresponding molding system (see Figure 4) with the runner system (see Figure 4a) and cooling system (see Figure 4b). Figure 4a shows that the PA66 melt was injected into the mold cavity with four gates through the runner system, and the size of the runners including the diameter and length is shown in the description. The coolant used in the entire injection molding system was water, and the water entered the cooling runner at a temperature of 25 °C to cool the mold (see Figure 4b). The injection molding process was then simulated using Moldex3D. The simulations considered five molding parameters, namely, the maximum injection pressure, the maximum packing pressure, the melt temperature, the mold temperature, and the cooling time. Each molding parameter (control factor) was assigned four different level settings. Hence, the Taguchi experiments were configured in an $L_{16}(5^4)$ orthogonal array (OA), as shown in Table 3 (gray region). The aim of the Taguchi experiments (i.e., Moldex3D simulations) was to determine the combination of control factor level settings which minimized the concentricity error of the BTL barrel at the position of each lens and hence optimized the coaxiality of the tube. Hence, for each run of the OA, the quality of the simulation outcome was evaluated using the following smaller-the-better S/N (signal-to-noise) ratio:

$$S/N = -10\log\left(\frac{1}{n}\sum_{i=1}^{n} y_i^2\right),\tag{1}$$

where y_i is the concentricity of the barrel at the measurement plane i ($i = 1\sim4$), and n is the number of measured points in the trial.

(a)

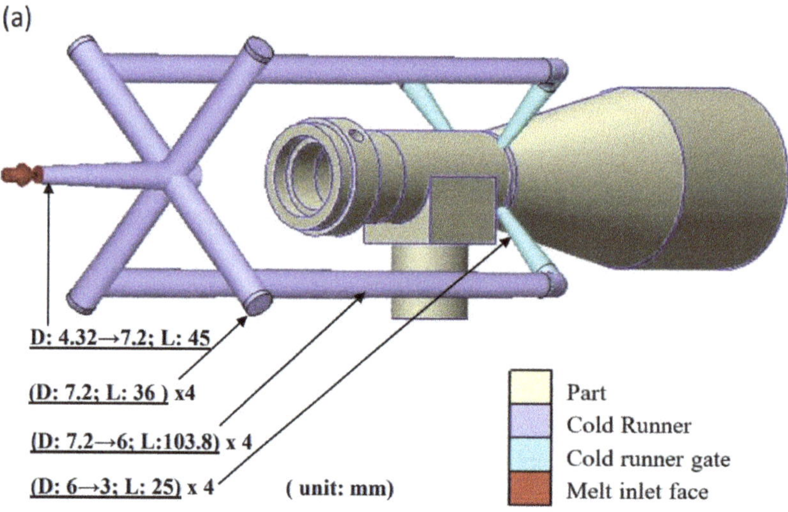

Figure 4. *Cont.*

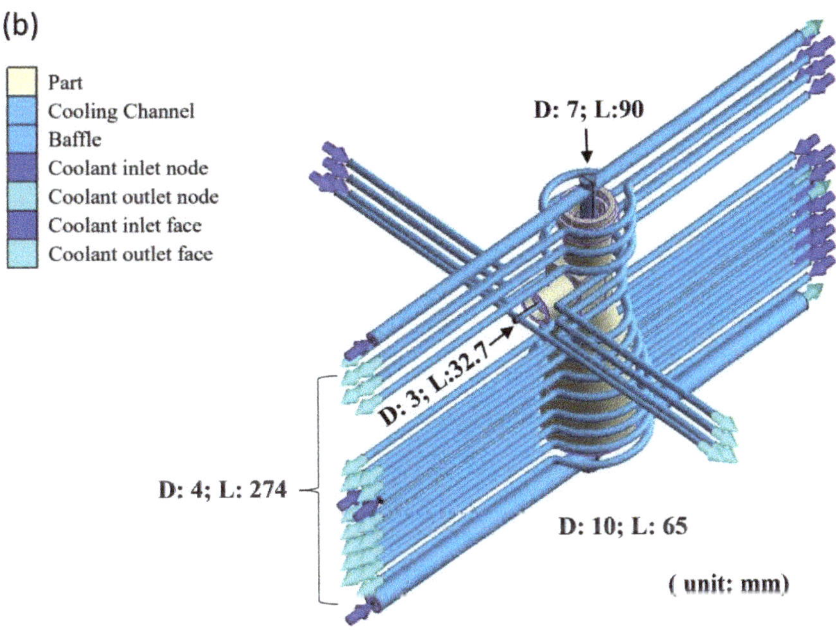

Figure 4. (**a**) Injection molding system and (**b**) cooling system of the BTL (D: diameter; L: length; unit: mm).

Table 3. Taguchi $L_{16}(5^4)$ orthogonal array for coaxiality optimization of the BTL.

	Processing Factors				
Trials	A Maximum Injection Pressure (MPa)	B Maximum Packing Pressure (MPa)	C Melt Temperature (°C)	D Mold Temperature (°C)	E Cooling Time (s)
General Parameters	200	200	280	80	13
1	180	180	275	70	11
2	180	200	280	80	13
3	180	220	285	90	15
4	180	240	290	100	17
5	200	180	280	90	17
6	200	200	275	100	15
7	200	220	290	70	13
8	200	240	285	80	11
9	220	180	285	100	13
10	220	200	290	90	11
11	220	220	275	80	17
12	220	240	280	70	15
13	240	180	290	80	15
14	240	200	285	70	17
15	240	220	280	100	11
16	240	240	275	90	13
Optimal Parameters	220	240	275	90	17

2.3. Roundness, Concentricity, and Coaxiality Measures

For cylindrical objects such as the BTL barrel considered in the present study, the manufacturing quality can be evaluated by three different metrics: the roundness, the concentricity, and the coaxiality. Traditional methods for evaluating the roundness of manufactured parts include the least squares circle (LSC) method, the minimum zone tolerance circle (MZC) method, the maximum inscribed circle (MIC) method, and the minimum circumscribed circle (MCC) method [19]. In the present study, the roundness of the BTL barrels produced using the different processing conditions specified in the Taguchi OA was evaluated using the LSC method [20]. In particular, a fitting process was applied to determine the circle which produced the minimum sum of the square error $F(x, y)$ with the actual circular profile of the tube, i.e.,

$$F(x,y) = \sum_{i=1}^{n} \left[r(x,y)^2 - R_c^2 \right]^2 = \sum_{i=1}^{n} \left[(x - x_c)^2 + (y - y_c)^2 - R_c^2 \right]^2 \tag{2}$$

where $r(x, y)$ is the distance between the measured point (x, y) and the unknown center (x_c, y_c) of the circle, n is the number of measured points, and R_c is the radius of the least squares circle. Taking the center (x_c, y_c) of the least squares error (LSE) circle as the center of the circle after processing deformation, the LSE circle center was calculated at each of the four reference planes (i.e., lens positions) in the BTL barrel. The concentricity, d, of the barrel at each plane was then computed as the offset of the LSE circle center from the designed datum value $(x_o, y_o) = (0, 0)$. Finally, the concentricity values obtained at the four reference planes were averaged to evaluate the coaxiality of the tube [21–23]. For convenience, the corresponding mathematical derivations are given in the following.

The concentricity of the BTL at plane i was evaluated as

$$d_i = \sqrt{(x_c - x_o)^2 + (y_c - y_o)^2} \tag{3}$$

The coaxiality (overall concentricity) of the BTL was computed as

$$\text{Coaxiality} = \sum_{i=1}^{n} [d_i]^2 / n \tag{4}$$

2.4. CODE V Optical Evaluation Using Eccentric Coordinates (x_c, y_c, z_c)

Following the Taguchi/Moldex3D simulations, the concentricity coordinates (x_c, y_c) and axial displacements (z_c) of each lens in the barrel were input to CODE V software in order to evaluate the optical performance of the device. The performance was evaluated in terms of three metrics, namely, the MTF, the spot diagram, and the 2D imaging simulation. In general, the MTF indicates how a lens reproduces contrast over a broad range of spatial frequencies, while the spot diagram computes the focused image size and provides a quick and qualitative assessment of the overall quality of the lens assembly. Finally, the 2D imaging simulation (resolution results with grayscale analysis) provides a further simple visualization of the optical system performance [24].

3. Results and Discussion

3.1. Taguchi Optimization Results

The general (or standard for specific machines) processing parameters for the considered PA66 polymer material were taken from the manufacturer's specification as follows: maximum injection pressure 200 MPa, maximum packing pressure 200 MPa, melt temperature 280 °C, mold temperature 80 °C, and cooling time 13 s (see the upper row in Table 3). However, the Taguchi analysis results indicated that the optimal parameter settings (i.e., the parameter settings which minimized the coaxiality error of the BTL tube) were a maximum injection pressure of 220 MPa, a maximum packing pressure of 240 MPa, a melt temperature of 275 °C, a mold temperature of 90 °C, and a cooling time of 17 s (see the lower row in Table 3). Table 4 shows the Taguchi S/N response data for the five control factors and their level settings. As shown, the control factors can be ranked in terms of a diminishing effect on the coaxiality of the tube as follows: maximum packing pressure > melt temperature > cooling time > maximum injection pressure > mold temperature. In other words, the coaxiality of the molded tube is dominated by the packing pressure and melt temperature, while the cooling time, injection pressure, and mold temperature all have more minor effects. Table 4 shows that the S/N ranges of factor A (maximum injection pressure) and factor D (molding temperature) are much smaller than those of factor B (maximum packing pressure), factor C (melt temperature), and factor E (cooling time). It is worth noting that under the condition of coaxial optimization of control factors A and D, their individual influence is significantly smaller than that of other factors (B, C, and E). However, factor A and factor D are different in the S/N range (but their values are very close). Therefore, their impact evaluation should be discussed in consideration of other different processing conditions.

Table 4. Signal-to-noise (S/N) response of control factors and level settings.

Factor Level/ Range/Rank	Control Factors				
	A	**B**	**C**	**D**	**E**
Level 1 (dB)	18.42426	17.93074	18.64934	18.4332	18.38978
Level 2 (dB)	18.44004	18.29598	18.51562	18.45852	18.36406
Level 3 (dB)	18.4714	18.63066	18.3822	18.45247	18.44444
Level 4 (dB)	18.4491	18.92741	18.23765	18.44061	18.58652
S/N Range (dB)	0.047142	0.996663	0.41169	0.025322	0.22246
Influence Rank	4	1	2	5	3

In general, the results show that a larger maximum packing pressure, a lower melt temperature, and a longer cooling time are beneficial in reducing the concentricity error (in each plane) of the tube and improving the coaxiality as a result. Table 5 shows the injection molding simulation parameters for the general design and the optimal design to verify the results of the Taguchi method.

Figure 5 shows the real sprue pressure changes and real flow rate changes in the filling stage and the packing stage after the injection molding simulation. It can be seen that increasing the maximum packing pressure and the maximum injection pressure helps to maintain the stability of the flow rate (about 31.97 cm^3/s) and is very close to the expected constant flow rate (34.05 cm^3/s). For the V/P (velocity-to-pressure) transition point, the general design and optimal design have values of 2.098 s and 2.097 s, respectively. The setting of the maximum injection pressure helps to keep the flow rate at a certain level during the filling stage and limits the injection pressure during the filling stage to prevent mold or machine failure. In fact, the real injection pressure also changes at the set desired flow rate, and the real injection pressure is always less than the maximum injection pressure during the filling stage. On the other hand, an injection molding machine with such a high maximum packing pressure setting (200 MPa) is an extreme upper limit and is not widely used throughout the process. Because this paper is a numerical simulation analysis, the setting of the packing pressure does not fully consider the actual situation. Generally speaking, when the packing pressure is too high, the pressure in the mold will be too large. Such a processing procedure can cause stress concentration and residual stress, which can lead to difficulty in ejection and even cracking. For the peak flow rate shown in Figure 5, this is not a typical velocity profile for injection molding. The possible reason is that the numerical simulation in this study adopted a one-stage flow rate setting. Under the influence of all the complex processing factors, the cavity space that is not fully filled must be filled before the expected end of time. Therefore, there is a tendency for the flow rate to suddenly increase before entering the packing stage. Of course, this situation still needs further confirmation and discussion.

Table 5. Injection molding process parameters for general design and optimal design.

	General Design	Optimal Design
Filling		
Filling time (s)	2.13	2.13
Melt temperature (°C)	280	275
Mold temperature (°C)	80	90
Maximum injection pressure (MPa)	200	220
Injection volume (cm^3)	68.1	68.1
Packing		
Packing time (s)	7	7
Maximum packing pressure (MPa)	200	240
Cooling		
Cooling time (s)	13	17
Mold-open time (s)	5	5
Eject temperature (°C)	190	190
Air temperature (°C)	25	25
Miscellaneous		
Cycle time (s)	27.13	27.13

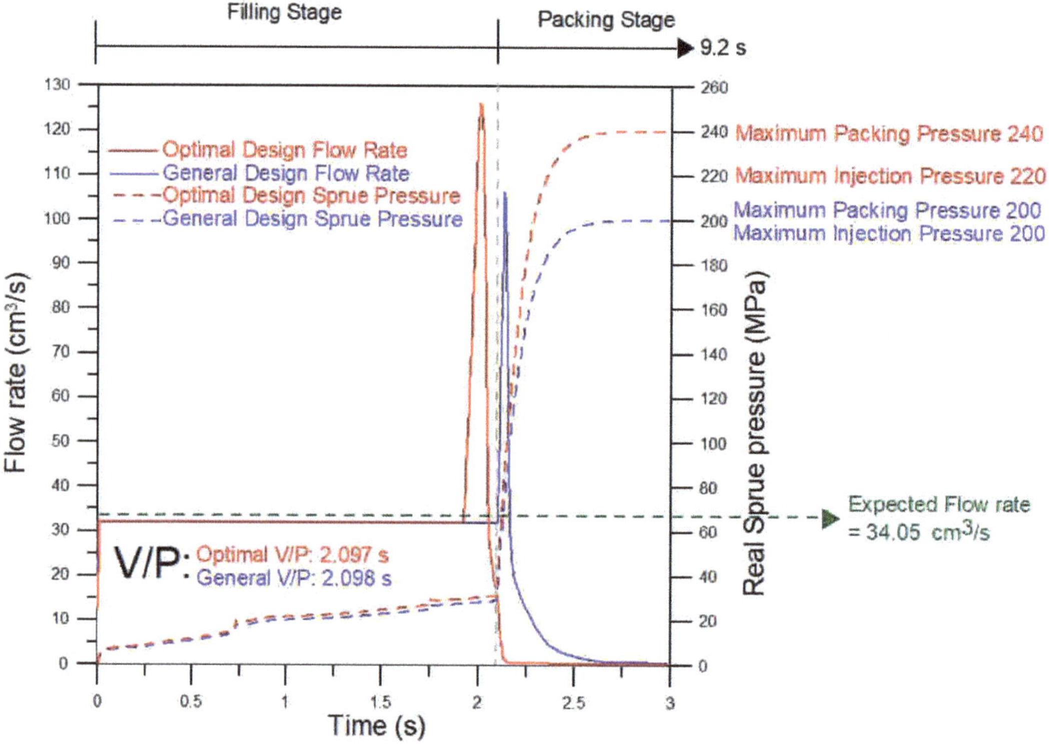

Figure 5. Considering the processing factors (maximum injection pressure and maximum packing pressure), the real sprue pressure change and flow rate change presented by the general design and the optimal design during the filling stage and the packing stage are shown.

Figure 6 plots the coaxiality error and S/N values for each of the 16 runs in the $L_{16}(5^4)$ OA, together with those for the general (i.e., recommended) processing parameters and optimal processing parameters. As shown, the tube produced using the optimal molding parameters has the maximum S/N value of 19.33078 dB and the minimum coaxiality error of 0.011666 mm^2. The S/N value and coaxiality error of the optimal tube are 5.76% higher and 21.5% lower, respectively, than those of the tube produced using the recommended processing parameters. In other words, the effectiveness of the optimized parameter settings in improving the coaxiality of the BTL tube is confirmed.

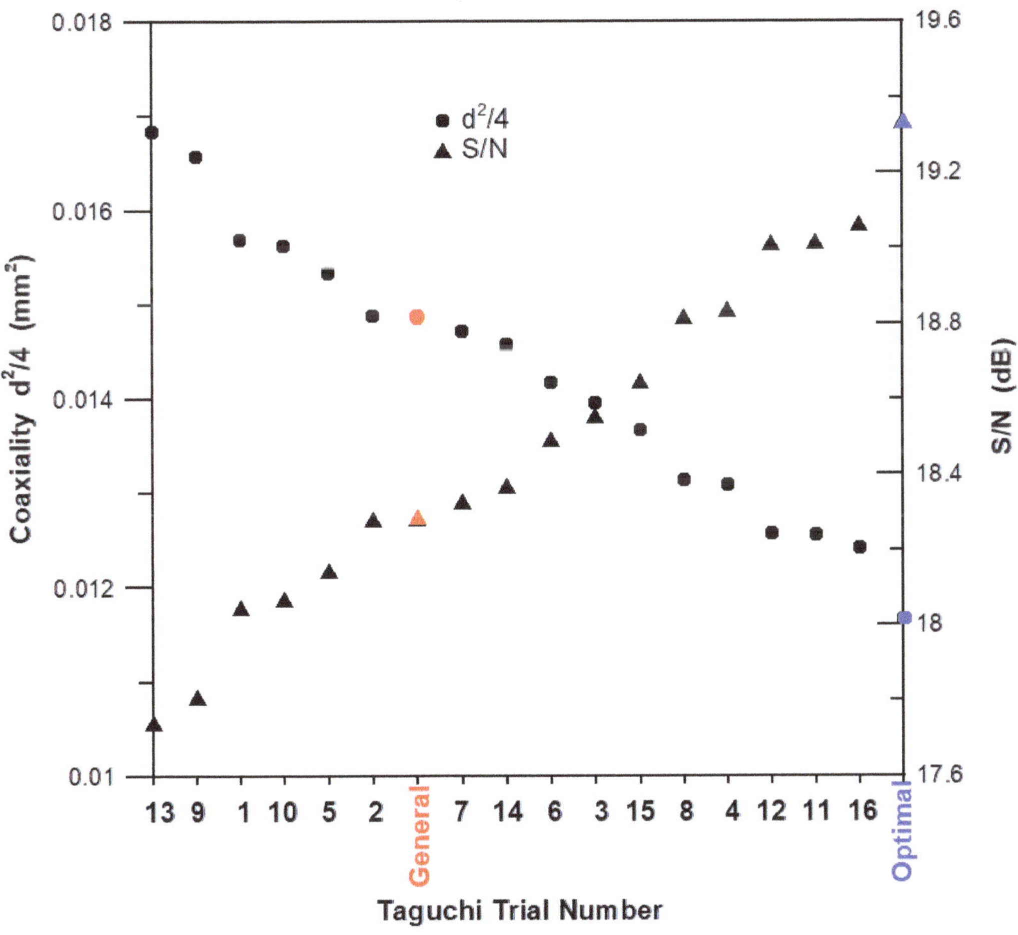

Figure 6. Coaxiality and S/N values for different control factor level settings in Moldex3D simulations.

3.2. Geometry Analysis

Table 6 and Figure 7 show the center coordinate positions and concentricity values at the four reference planes (Z_1~Z_4) of the ideal BTL design (with zero concentricity and coaxiality errors), and the BTL tubes produced using the general and optimal processing conditions. The results indicate that the geometric distortion produced by the manufacturing process is far less than that produced by the asymmetric design of the barrel itself (i.e., the protruding handle structure (see Figure 1)). In addition, the optimal process parameters have a greater effect in reducing the concentricity error at measurement planes Z_1 and Z_4 (at the two ends of the BTL barrel) than at planes Z_2 and Z_3 (in the central region of the barrel). Notably, however, the results presented in Figure 7 show that even though the concentricity values at the four planes are not markedly improved in the optimal design, the overall coaxiality of the tube is significantly improved compared to that of the tube produced using the general parameters (i.e., from 0.014866 to 0.011666 mm, as shown in Table 6). It should be noted that the runner design and cooling system in this study are based on virtual simulation. If the injection molding process is to be carried out realistically, the interference problem caused by the mold assembly must be considered.

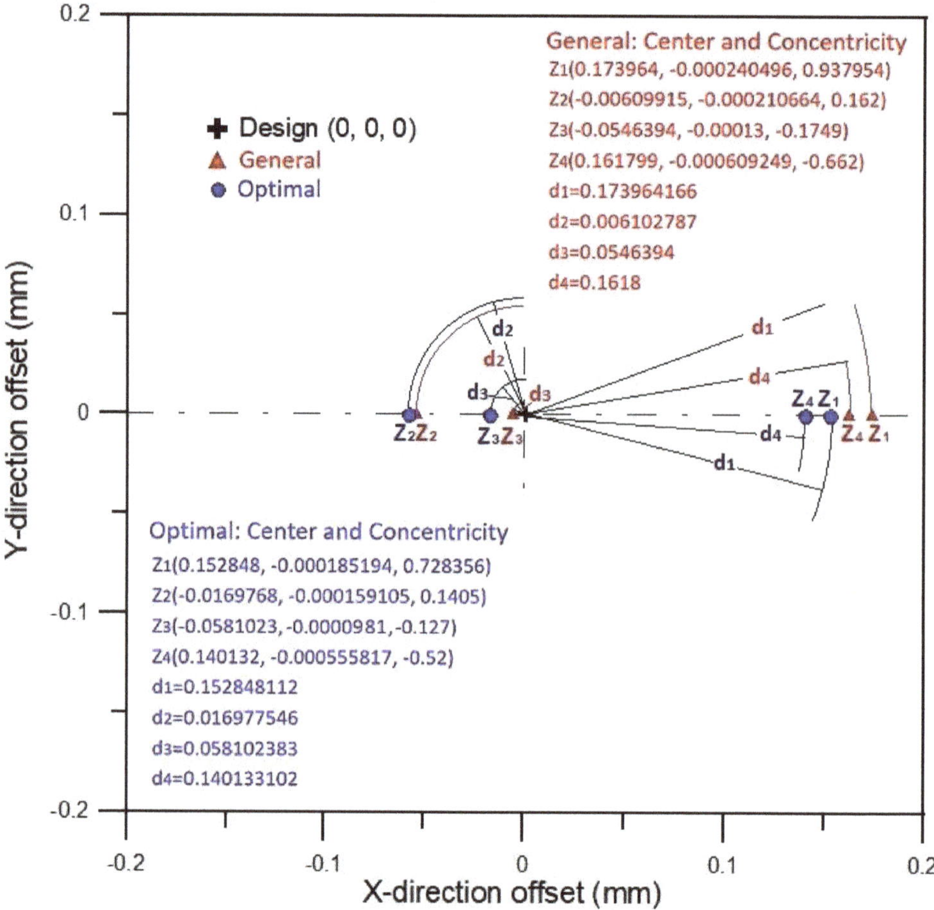

Figure 7. Center distribution and concentricity values in the design, general, and optimal BTL models.

Table 6. Center coordinates/concentricity values of four reference planes and coaxiality of the optical tube in the design, general, and optimal BTL models.

Unit: mm	Design	General	Optimal
Planes		Center Coordinate Bias : $Z_i(\Delta Xc,\ \Delta Yc,\ \Delta Zc)$; Concentricity $= d_i\ (i = 1\text{–}4)\,(\text{mm})$; Coaxiality $= \sum_{i=1}^{4}(d_i)^2/4\,(\text{mm}^2)$	
$Z1 = 0$	$(0, 0, 0)$ 0	Z1(0.173964, −0.000240496, 0.937954) $d1 = 0.173964166$	Z1(0.152848, −0.000185194, 0.728356) $d1 = 0.152848112$
$Z2 = 57.75$	$(0, 0, 0)$ 0	Z2(−0.00609915, −0.000210664, 0.162) $d2 = 0.006102787$	Z2(−0.0169768, −0.000159105, 0.1405) $d2 = 0.016977546$
$Z3 = 82.3$	$(0, 0, 0)$ 0	Z3(−0.0546394, −0.00013, −0.1749) $d3 = 0.0546394$	Z3(−0.0581023, −0.0000981, −0.127) $d3 = 0.058102383$
$Z4 = 117.7$	$(0, 0, 0)$ 0	Z4(0.161799, −0.000609249, −0.662) $d4 = 0.1618$	Z4(0.140132, −0.000555817, −0.52) $d4 = 0.140133102$
Coaxiality	0	0.014866	0.011666

3.3. Optical Performance

As described in Section 2.4, the optical performance of the BTL barrel produced using the optimal processing conditions was evaluated by means of MTF plots, spot diagrams, and 2D imaging simulations. For comparison purposes, CODE V simulations were also performed for the ideal design barrel and for the BTL barrel produced using the general processing conditions. Note that the center coordinates and concentricity values of the four lenses in each barrel were assigned the values shown in Table 6.

Figure 8 shows the sagittal and tangential MTF curves of the three BTL tubes (designated as design, general, and optimal) for light fields F_I, F_{II}, and F_{III} (see Figure 1). In general, an MTF curve closer to the diffraction limit curve indicates a better imaging performance. Consequently, all six MTF curves for the ideal BTL design (with zero concentricity errors) are virtually superimposed on the diffraction limit curve. Moreover, the MTF curves for the optimal tube are closer to the diffraction limit curve than those of the general tube. In other words, the optimized parameters yield a significant improvement in the optical performance of the BTL tube compared to that of the tube produced using the general processing parameters. Figures 9–11 provide a detailed analysis of the imaging performance of the three tubes for light fields I, II, and III. In general, the results confirm that, for all three light fields, the imaging performance of the optimal tube is closer to the ideal performance than that of the general tube. Furthermore, comparing the three figures, it is seen that the imaging performance of the general tube and optimal tube is close to that of the ideal tube for light rays close to the optical axis of the BTL (i.e., F_{III}) but gradually degrades with an increasing distance from the axis (i.e., F_{II} and F_I).

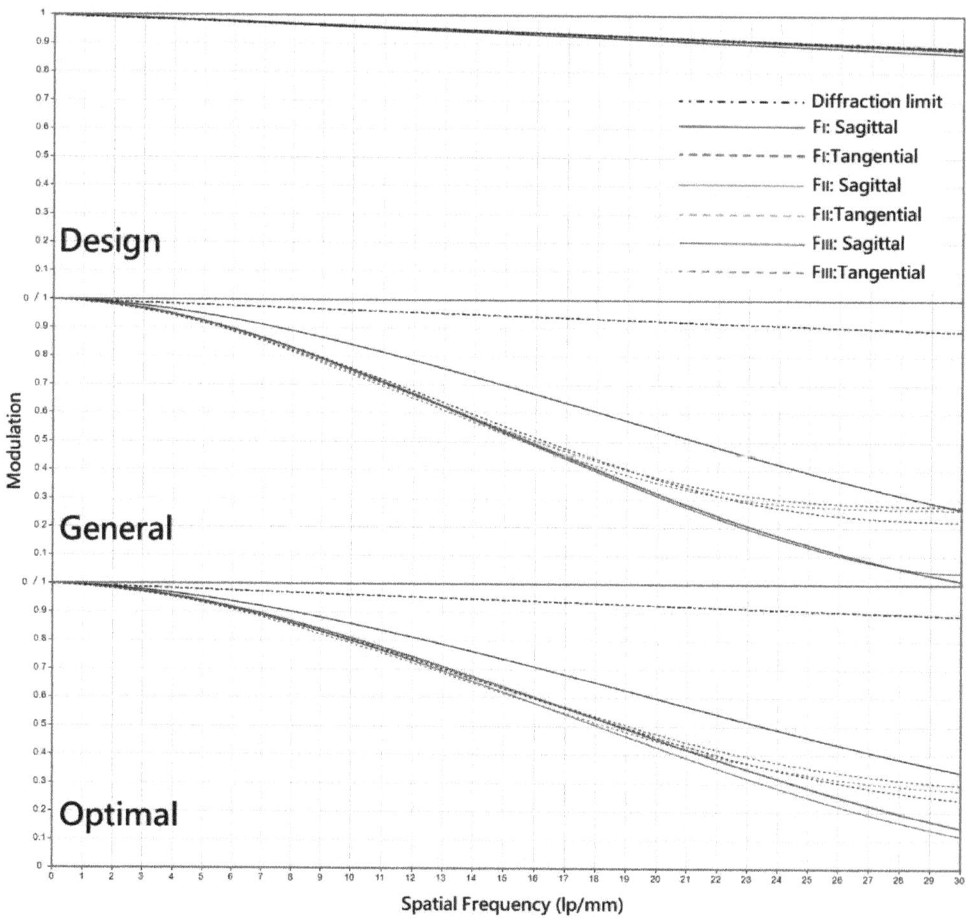

Figure 8. Modulation vs. spatial frequency response for different light fields in the design, general, and optimal BTL models.

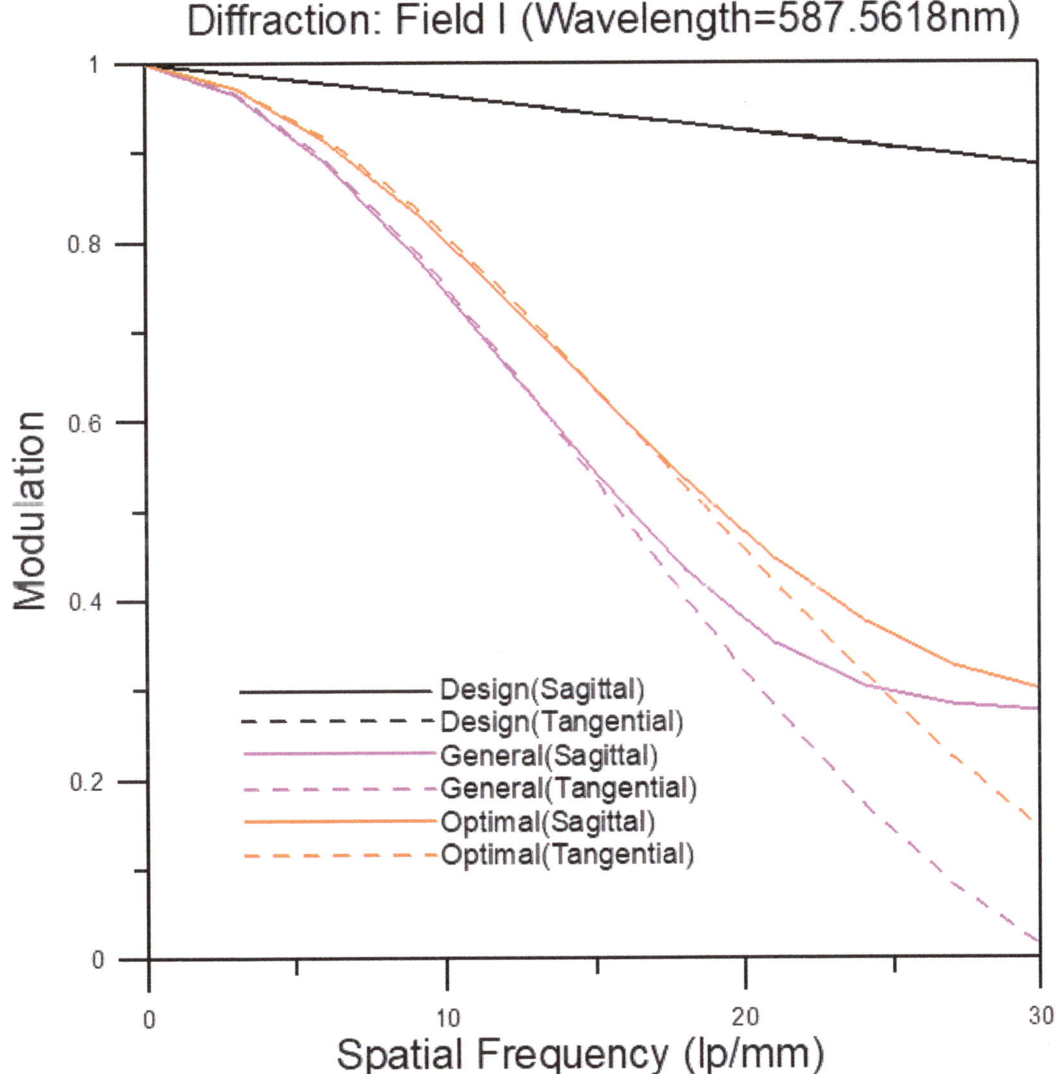

Figure 9. MTF for light field I in the design, general, and optimal BTL models.

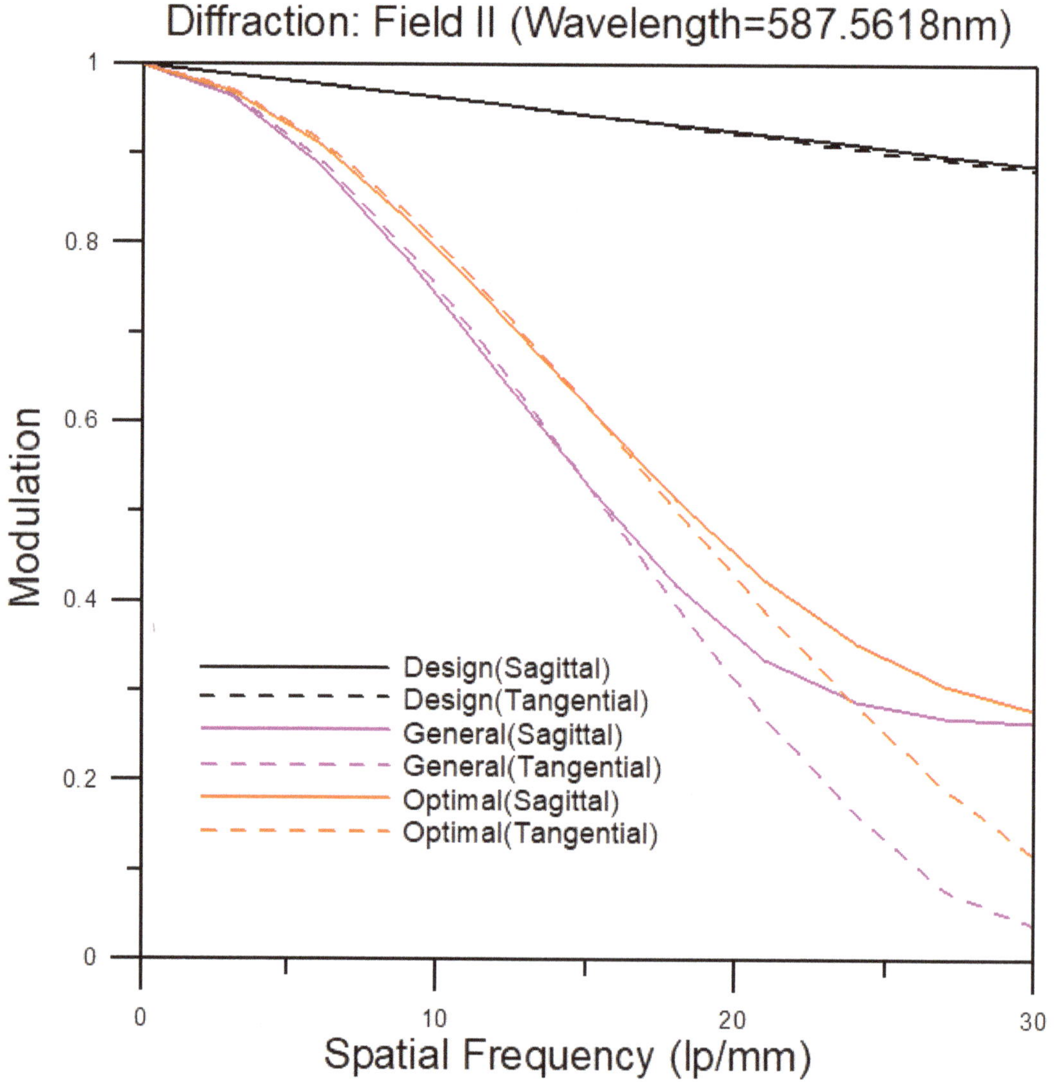

Figure 10. MTF for light field II in the design, general, and optimal BTL models.

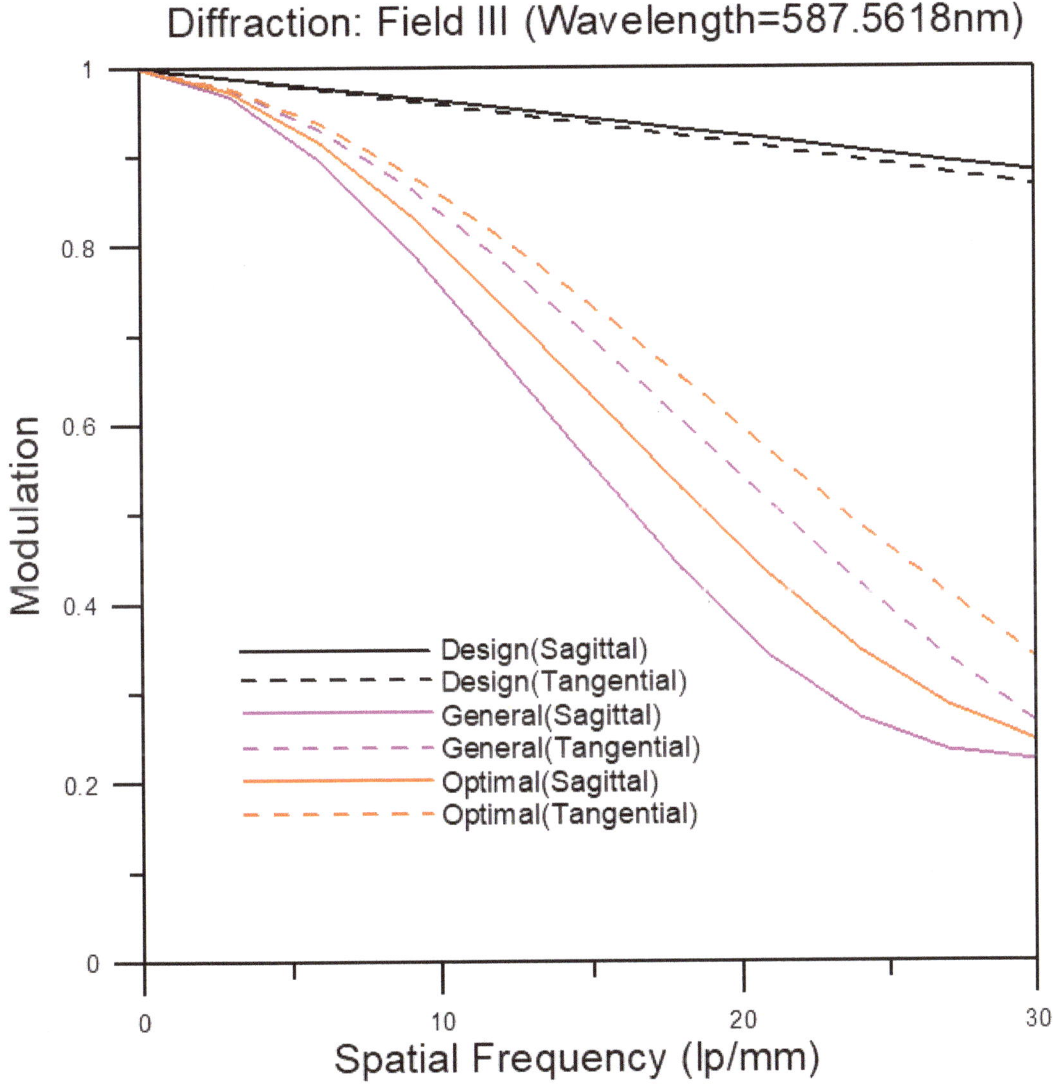

Figure 11. MTF for light field III in the design, general, and optimal BTL models.

Figure 12 shows the spot diagrams of the three light fields on the image plane of the design, general, and optimal BTL tubes. Note that for each spot, the RMS (root mean square) value indicates the root mean square values of all the light spot distribution coordinates, while the 100% value indicates the smallest circle diameter (in mm) sufficient to encompass all of the individual light spots within the light field. Observing the three sets of light spots for each tube, it is clear that the optimal tube results in a sharper imaging performance than the general tube. Furthermore, examining the spot diagrams for light field F_{III} (closest to the optical axis), the minimum circumscribed circle diameters of the design, general, and optimal tubes are found to be 0.002281 mm, 0.047270 mm, and 0.040569 mm, respectively. In other words, the optimal tube reduces the focused spot size by 14.2% compared to that of the general tube. The corresponding improvement values for the F_{II} and F_{III} fields are 12.8% and 10.5%, respectively. Similarly, the optimal tube improves the RMS value of the focus spot size by 15.4% compared to that of the general tube for light field F_{III}, and 13.7% and 11.1%, respectively, for light fields F_{II} and F_{I}. In other words, the spot diagram results confirm that the optimal tube, with an improved coaxiality, produces a better overall optical quality than the general tube.

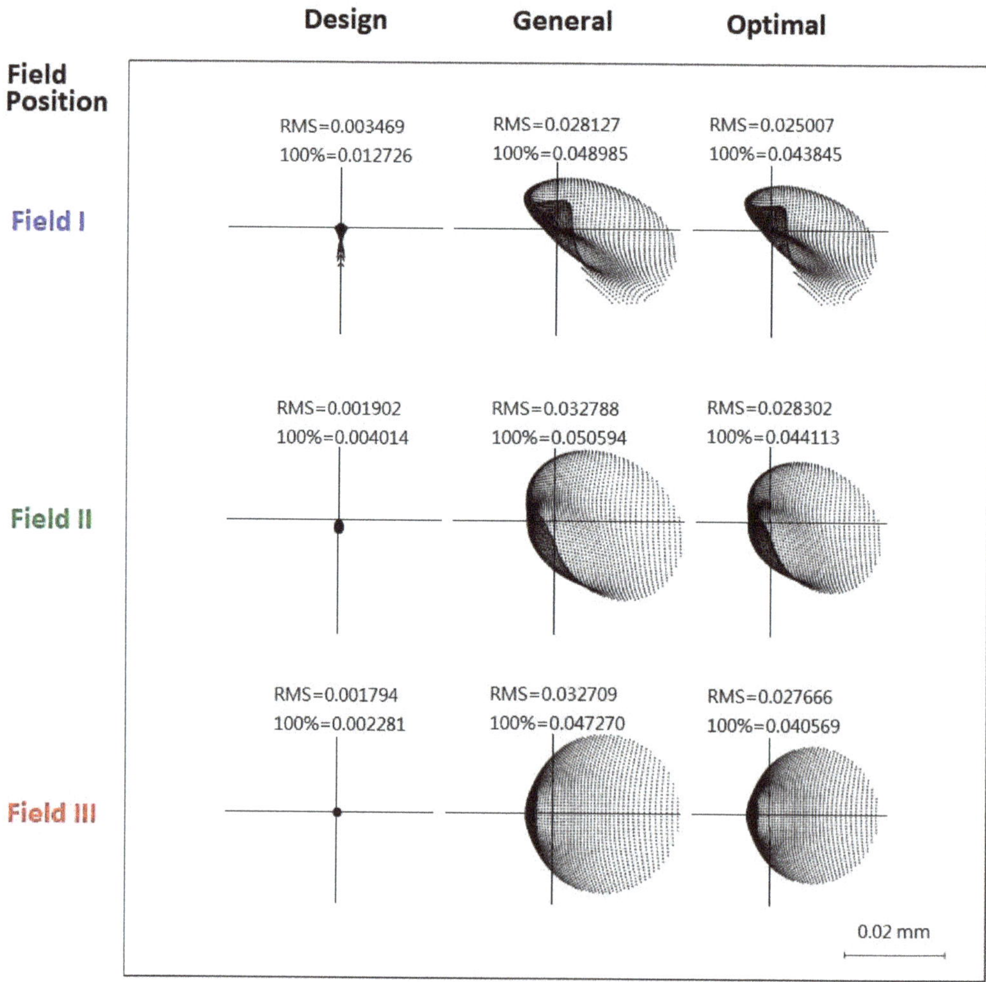

Figure 12. Spot diagrams of three light fields in the design, general, and optimal BTL models.

3.4. D Imaging Simulation: Grayscale Analysis of Imaging Quality

The imaging quality of the optimal BTL device was further evaluated by means of grayscale analyses. Figure 13 shows the 1951 USAF resolution test chart commonly used to evaluate the optical imaging quality of lens combinations [25]. Note that the dotted circles correspond to different fields of view (FOVs) with a gradually increasing scale from 0 (blue circle) to 1 (black circle). Figure 14 presents a qualitative comparison of the imaging performance of the ideal (design), general, and optimal BTL designs for the right-hand region of the resolution test chart and numeral "2" in the top-left corner of the chart. Although a slight improvement in the imaging quality is observed for the optimal design, it is unconvincing. Accordingly, quantitative grayscale analysis simulations were conducted along the yellow reference line indicated in the four images in Figure 14 and shown on the right-hand side of Figure 13. As shown in Figure 15a, the analysis was conducted over a distance of 340 pixels, extending from pixel 140 (prior to the first block of three vertical lines in the resolution test chart) to pixel 480 (after the final block of three vertical lines in the test chart). Figure 15b shows the variation in the grayscale intensity along the considered line as observed through the ideal BTL, general BTL, and optimal BTL. In Figure 15b, there are 18 peak groups corresponding to the black and white contrast area of the 18 areas crossed by the yellow line in Figure 15a. The black and white interface is grayscaled due to optical errors and imaging characteristics. These differences can be used to explain the pros and cons of imaging. To better evaluate and compare the imaging qualities of the three BTL designs, the grayscale profiles shown in Figure 15 were partitioned into three sub-profiles corresponding to the 1st peak group, 8th peak group, and 18th peak group in Figure 15. The corresponding results are presented in Figure 16a–c, respectively, where a grayscale curve closer to the objective curve indicates an improved imaging quality (i.e., a lower imaging distortion).

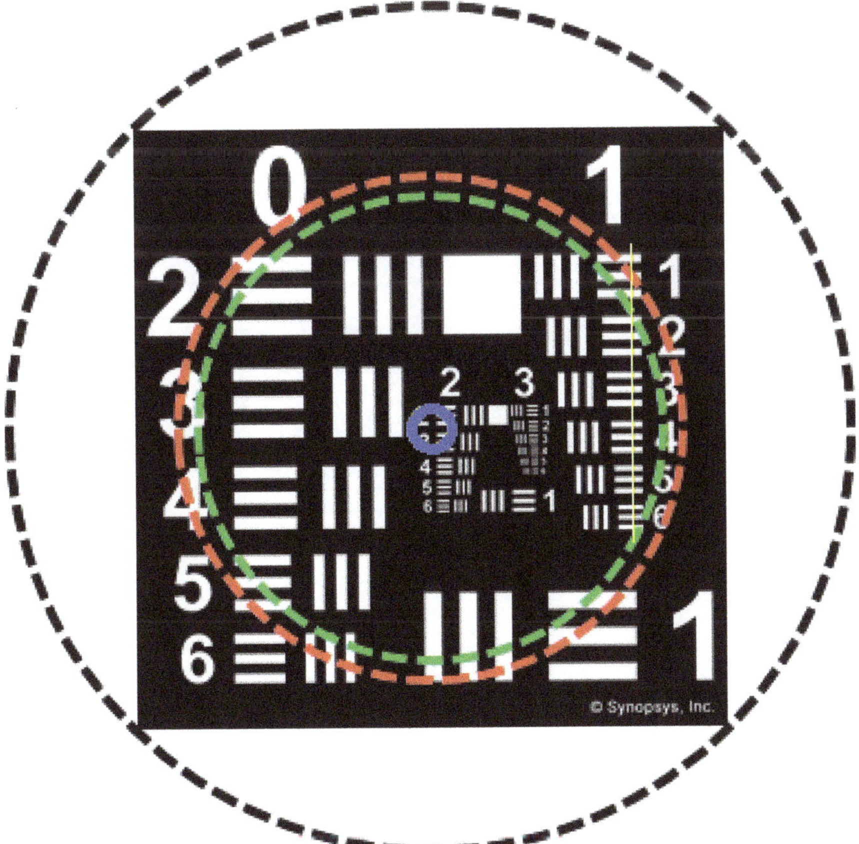

Figure 13. 1951 USAF resolution test chart with graduated FOV. Black circle: FOV = 1; red circle: FOV = 0.6; green circle: FOV = 0.55; blue circle: FOV = 0. Yellow line shows reference line used to examine image distortion and sharpness of different BTL designs by grayscale analysis.

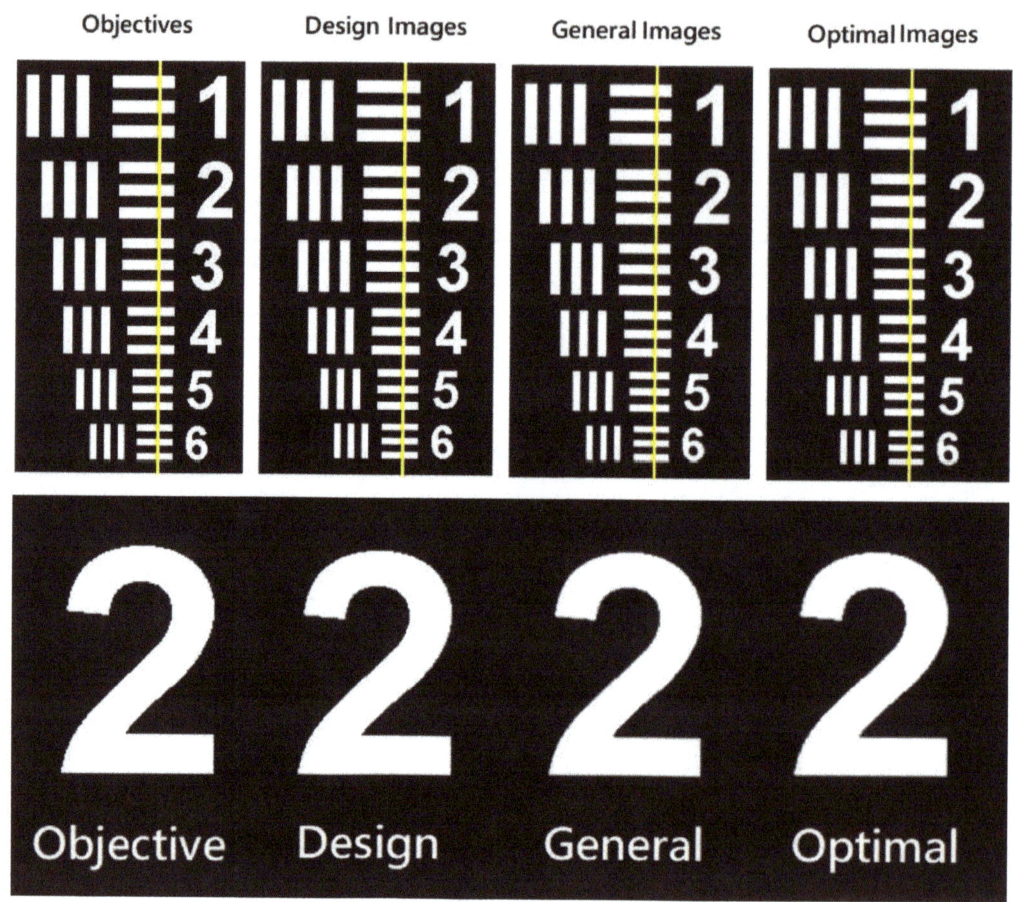

Figure 14. Two-dimensional image simulation results for the design, general, and optimal BTL models.

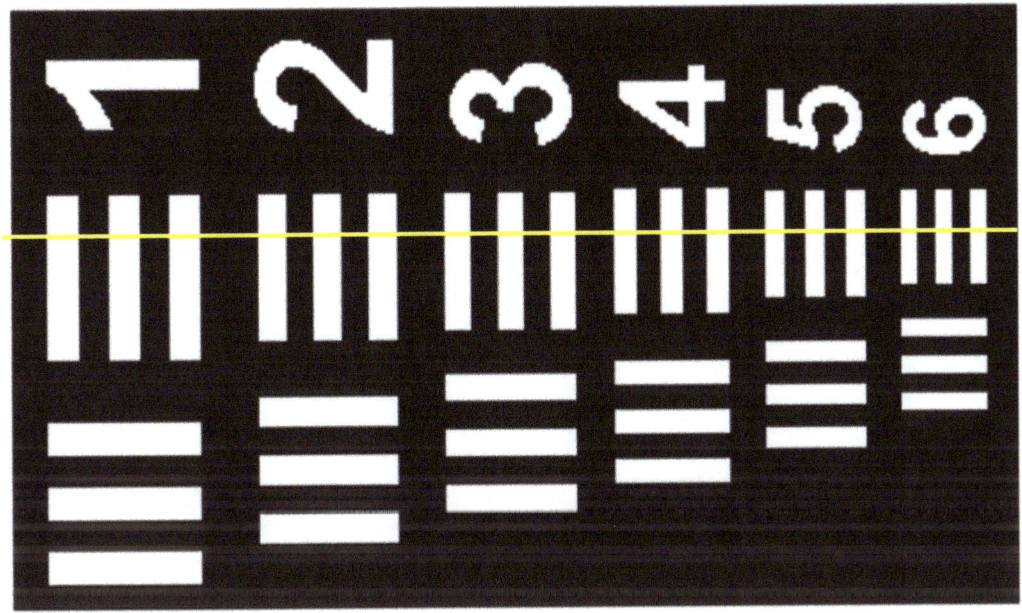

(**a**)

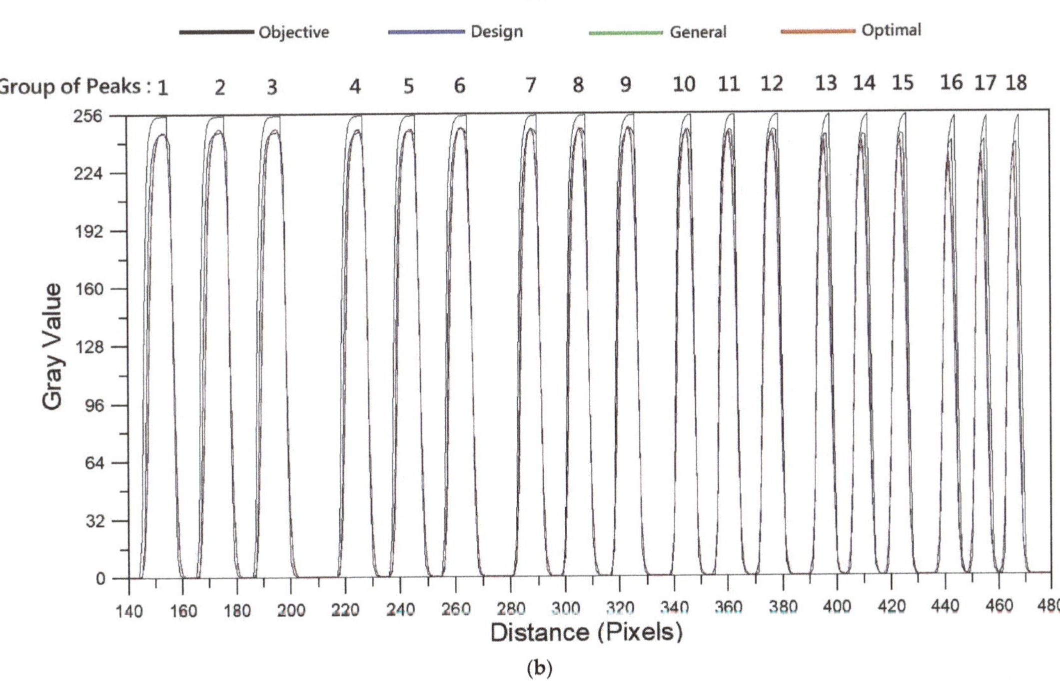

(**b**)

Figure 15. Quantitative image distortion and sharpness evaluation of objective, design, general, and optimal BTL models. (**a**) Analysis region in the USAF resolution test chart (refer to Figure 12, rotate 90 degrees counterclockwise). (**b**) Variation in grayscale intensity with pixel distance along yellow reference line.

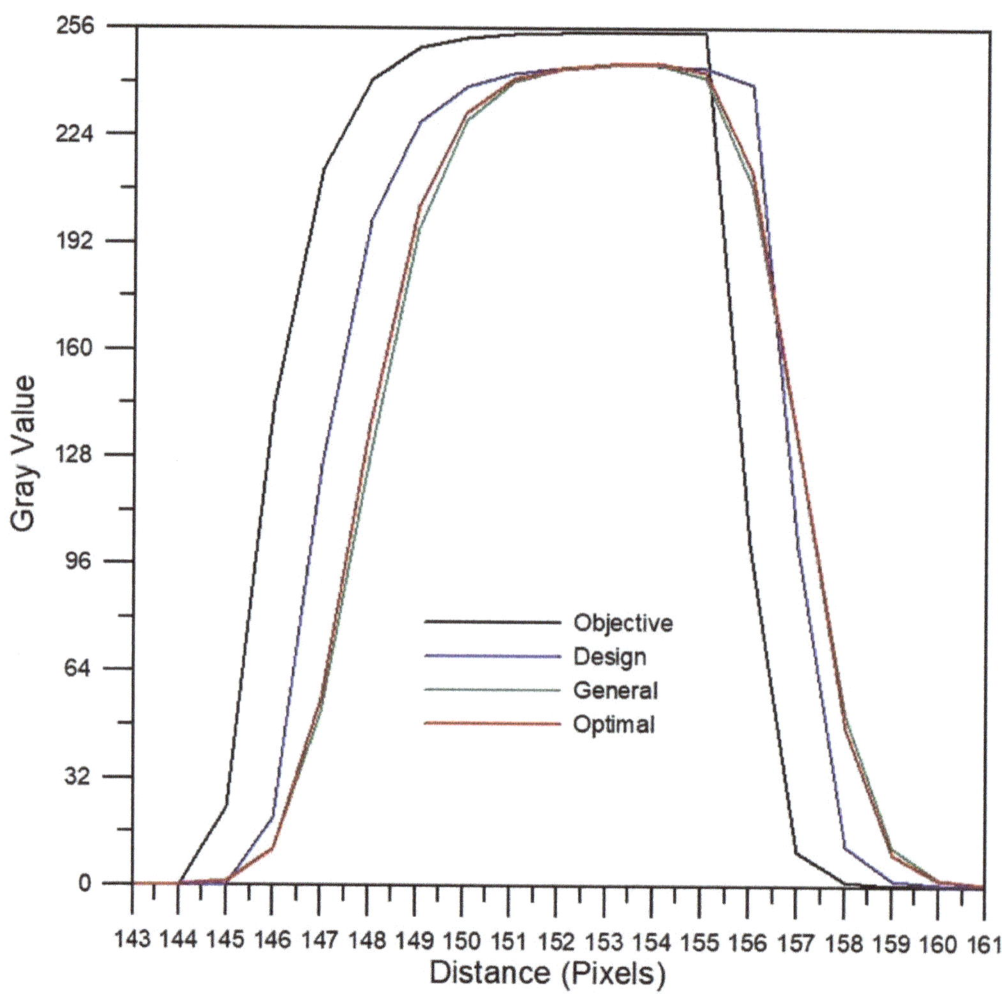

(**a**)

Figure 16. *Cont.*

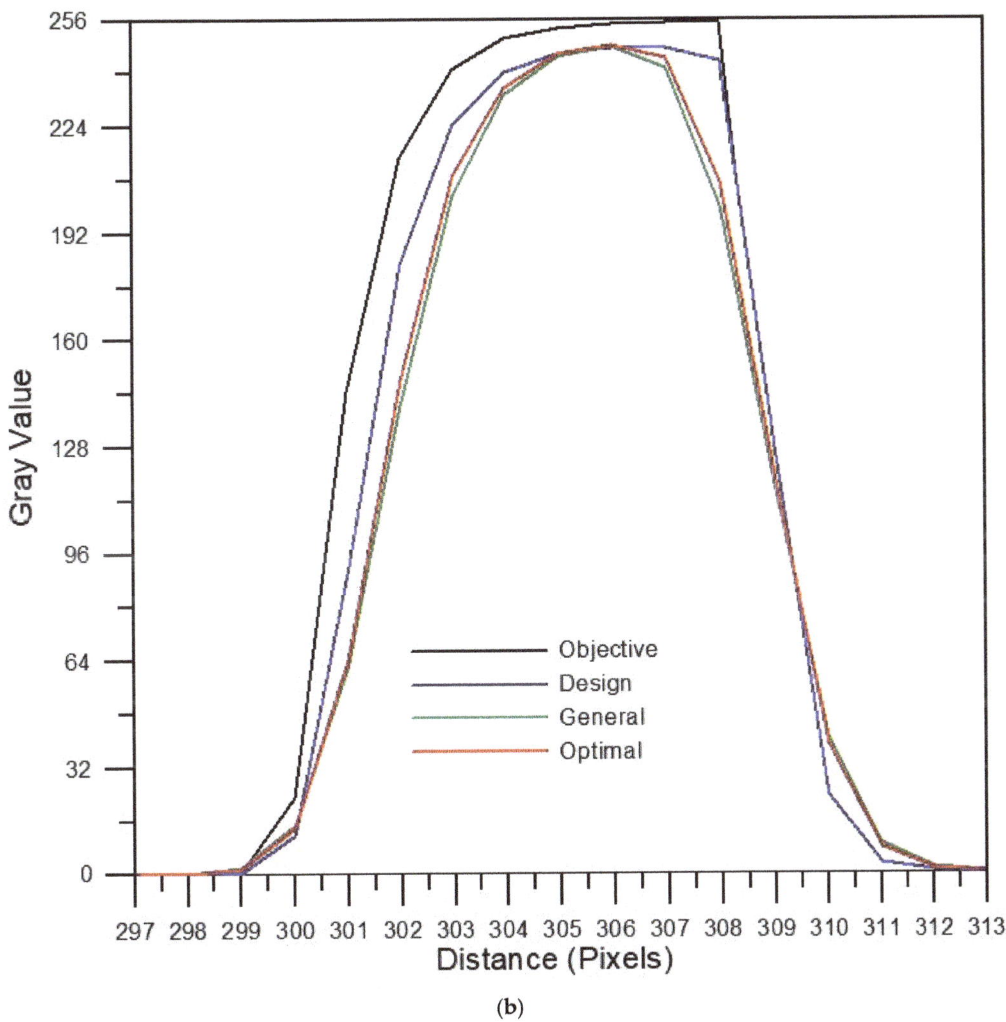

(**b**)

Figure 16. *Cont.*

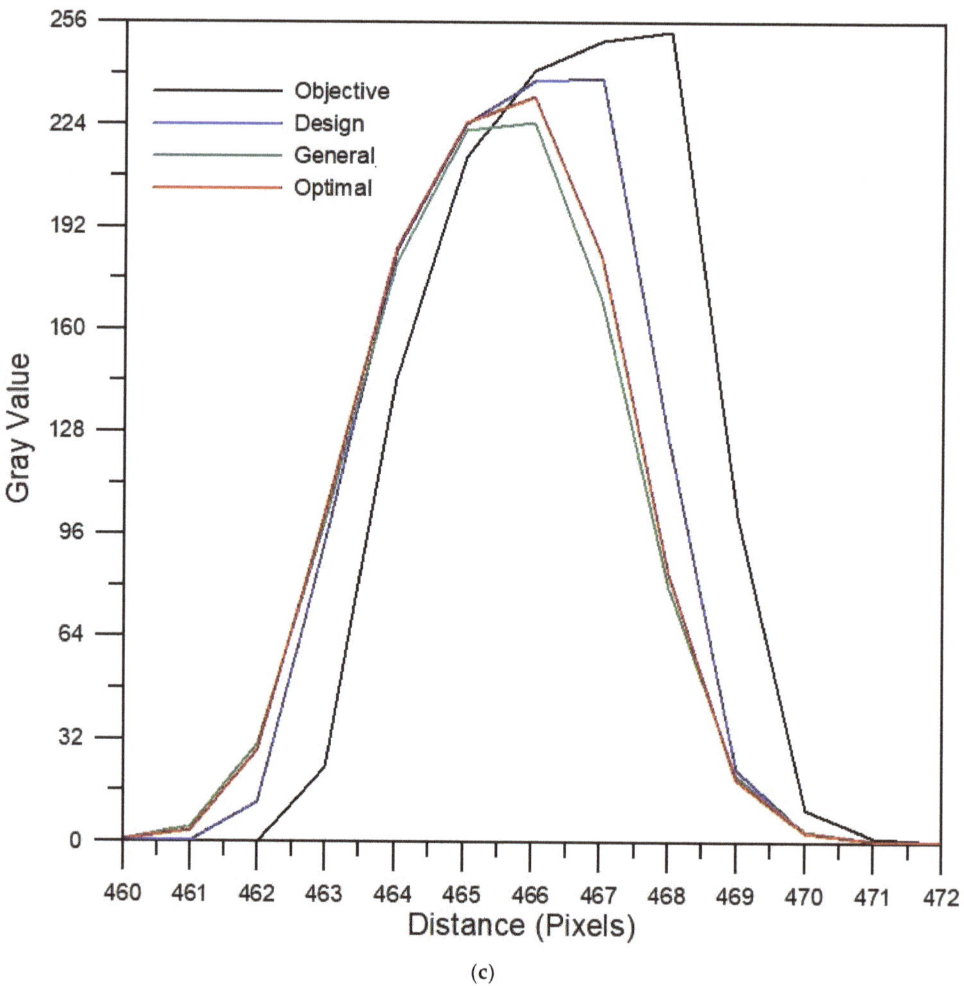

(c)

Figure 16. (**a**) Variation in grayscale intensity over pixel range of 143–161 (peak group 1 in Figure 15b). (**b**) Variation in grayscale intensity over pixel range of 297–313 (peak group 8 in Figure 15b). (**c**) Variation in grayscale intensity over pixel range of 460–472 (peak group 18 in Figure 15b).

Figure 16 shows that the 1st peak group is the imaging section roughly between pixel distances 143 and 161, the 8th peak group is the imaging section roughly between pixel distances 297 and 313, and the 18th peak group is the imaging section roughly between pixel distances 460 and 472. Observing the three figures, it is clear that the design curve (i.e., the ideal BTL curve) lies closest to the objective curve in all three cases. Moreover, the optimal curve lies closer to the objective curve than the general curve in every case. In other words, the superior imaging quality of the optimized BTL design compared to the general BTL design is confirmed. For the grayscale curves shown in Figure 16a–c, a steeper gradient in the interface regions between the black and white features in the resolution test image implies a greater imaging sharpness. Thus, a detailed inspection of the grayscale analysis results for the optimal and general BTL designs confirms that the optimal design yields a better image sharpness.

The grayscale curves shown in Figure 16a–c correspond to the upper-right, central-right, and lower-right regions of the objective image, respectively (see Figure 13). Comparing the grayscale curves in the three figures, it is seen that the image distortions of the three BTL designs in Figure 16a,c are relatively large and are located to the upper right and lower right of the objective image (see Figure 13), respectively. This result is reasonable since the FOVs are quite large in both cases (i.e., ~0.6 and 0.55, respectively) and the target geometry is not perpendicular to the direction of the FOV, resulting in a larger distortion and asymmetry, respectively. However, the grayscale curves in Figure 16b show both a smaller distortion and a greater symmetry since the FOV is smaller (i.e., 0.45 to 0.5) and the target geometry is perpendicular to the direction of the FOV.

4. Conclusions

This study combined CODE V optical analysis software, the Taguchi experimental design method, and Moldex3D flow analysis simulations to optimize the injection molding processing parameters for a bilateral telecentric lens (BTL) barrel in such a way as to maximize the coaxiality of the tube and improve the imaging performance as a result. The Taguchi analysis results show that a larger maximum packing pressure, a lower melt temperature, and a longer cooling time are all beneficial in improving the coaxiality of the BTL barrel. Moreover, the CODE V and grayscale analysis results show that the BTL barrel molded using the optimal processing conditions yields a better MTF response and focusing performance than a BTL tube molded using the general processing parameters for the chosen polymer material (PA66).

Author Contributions: Conceptualization, C.-M.L., Y.-J.C.; formal analysis, C.-M.L., Y.-J.C.; writing—original draft preparation, C.-M.L., Y.-J.C.; data curation, Y.-J.C.; writing—review and editing, C.-M.L.; supervision, C.-M.L.; funding acquisition, C.-M.L. All authors have read and agreed to the published version of the manuscript.

Funding: This research was funded by the Ministry of Science and Technology of Taiwan, ROC, under Grant Numbers MOST 109-2221-E-415-001-MY3 and MOST 109-2221-E-415-002-MY3.

Institutional Review Board Statement: Not applicable.

Informed Consent Statement: Not applicable.

Acknowledgments: The authors gratefully acknowledge the financial support provided to this study by the Ministry of Science and Technology of Taiwan, ROC.

Conflicts of Interest: The authors declare no conflict of interest.

References

1. Lin, C.M.; Tan, C.M.; Wang, C.K. Gate design optimization in the injection molding of the optical lens. *Optoelectron. Adv. Mater. Rapid Commun.* **2013**, *7*, 580–584.
2. Lin, C.M.; Hsieh, H.K. Processing optimization of Fresnel lenses manufacturing in the injection molding considering birefringence effect. *Microsyst. Technol.* **2017**, *23*, 5689–5695. [CrossRef]
3. Lin, C.M.; Chen, Y.W. Grey optimization of injection molding processing of plastic optical lens based on joint consideration of aberration and birefringence effects. *Microsyst. Technol.* **2019**, *25*, 621–631. [CrossRef]
4. Lu, X.; Khim, L.S. A statistical experimental study of the injection molding of optical lenses. *J. Mater. Process. Technol.* **2001**, *113*, 189. [CrossRef]
5. Lee, Y.B.; Kwon, T.H.; Yoon, K. Numerical prediction of residual stresses and birefringence in injection/compression molded center-gated disk. Part I: Basic modeling and results for injection molding. *Polym. Eng. Sci.* **2002**, *42*, 2246–2272. [CrossRef]
6. Shyu, G.D.; Isayyev, A.I.; Lee, H.S. Numerical simulation of flow-induced birefringence in injection molded disk. *Jpn. Korea Plast Process Jt. Semin.* **2003**, *4*, 41–47.
7. Fan, B.; Kazmer, D.O.; Bushko, W.C.; Theriault, R.P.; Poslinski, A.J. Birefringence prediction of optical media. *Polym. Eng. Sci.* **2004**, *44*, 814–824. [CrossRef]
8. Liao, S.J.; Chang, D.Y.; Chen, H.J.; Tsou, L.S.; Ho, J.R.; Yau, H.T.; Hsieh, W.H.; Wang, J.T.; Su, Y.C. Optimal process conditions of shrinkage and warpage of thin-wall parts. *Polym. Eng. Sci.* **2004**, *44*, 917–928. [CrossRef]
9. Tang, S.H.; Tan, Y.J.; Sapuan, S.M.; Sulaiman, S.; Ismail, N.; Samin, R. The use of Taguchi method in the design of plastic injection mould for reducing warpage. *J. Mater. Process. Technol.* **2007**, *182*, 418–426. [CrossRef]
10. Bociaga, E.; Jaruga, T.; Lubczynska, K.; Gnatowski, A. Warpage of injection moulded parts as the result of mould temperature difference. *Arch. Mat. Sci. Eng.* **2010**, *44*, 28–34.

11. Shayfull, Z.; Ghazali, M.F.; Azaman, M.; Nasir, S.M.; Faris, N.A. Effect of differences core and cavity temperature on injection molded part and reducing the warpage by Taguchi method. *Int. J. Eng. Technol.* **2010**, *10*, 125–132.
12. Fischer, R.E.; Tadic-Galeb, B.; Yoder, P.R. *Optical System Design*, 2nd ed.; McGraw-Hill: New York, NY, USA, 2008; pp. 179–198.
13. Dilworth, D.C. New tools for the lens designer, Current Developments in Lens Design and Optical Engineering IX. *Int. Soc. Opt. Photonics* **2008**, *7060*, 70600B.
14. Sahin, F.E. Lens design for active alignment of mobile phone cameras. *Opt. Eng.* **2017**, *56*, 065102. [CrossRef]
15. Grabarnik, S. Optical design method for minimization of ghost stray light intensity. *Appl. Opt.* **2015**, *54*, 3083–3089. [CrossRef]
16. Sahin, F.E. Open-source optimization algorithms for optical design. *Optik* **2019**, *178*, 1016–1022. [CrossRef]
17. Houllier, T.; Lépine, T. Comparing optimization algorithms for conventional and freeform optical design: Erratum. *Opt. Express* **2019**, *27*, 28383. [CrossRef]
18. Lin, C.M.; Chen, Y.J. Taguchi Optimization of Roundness and Concentricity of a Plastic Injection Molded Barrel of a Telecentric Lens. *Polymers* **2021**, *13*, 3419. [CrossRef]
19. Sui, W.; Zhang, D. Four Methods for Roundness Evaluation. *Phys. Procedia* **2012**, *24*, 2159–2164. [CrossRef]
20. Ahn, S.J. *Least Squares Orthogonal Distance Fitting of Curves and Surfaces in Space*; Springer: Berlin/Heidelberg, Germany, 2004.
21. Srinivasan, V.; Shakarji, C.M.; Morse, E.P. On the enduring appeal of least-squares fitting in computational coordinate metrology. *J. Comput. Inf. Sci. Eng.* **2011**, *12*, 011008. [CrossRef]
22. Kim, N.H.; Kim, S.W. Geometrical tolerances: Improved linear approximation of least-squares evaluation of circularity by minimum variance International. *J. Mach. Tools Manuf.* **1996**, *36*, 355–366. [CrossRef]
23. Hichem, N.; Pierre, B. Evaluation of roundness error using a new method based on a small displacement screw. *Meas. Sci. Technol.* **2014**, *25*, 25.
24. O'Shea, D.C.; Bentley, J.L. *Designing Optics Using CODE V*; SPIE PRESS: Bellingham, WA, USA, 2018.
25. *MIL-STD-150A*; Military Standard: Photographic Lenses. EverySpec LLC: Gibsonia, PA, USA, 12 May 1959.

Article

An Approximate Solution for the Contact Problem of Profiles Slightly Deviating from Axial Symmetry

Valentin L. Popov [1,2]

1   Department of System Dynamics and Friction Physics, Technische Universität Berlin, 10623 Berlin, Germany; v.popov@tu-berlin.de
2   National Research Tomsk State University, 634050 Tomsk, Russia

Abstract: An approximate solution for a contact problem of profiles which are not axially symmetrical but deviate only slightly from the axial symmetry is found in a closed explicit analytical form. The solution is based on Betti's reciprocity theorem, first applied to contact problems by R.T. Shield in 1967, in connection with the extremal principle for the contact force found by J.R. Barber in 1974 and Fabrikant's approximation (1986) for the pressure distribution under a flat punch with arbitrary cross-section. The general solution is validated by comparison with the Hertzian solution for the contact of ellipsoids with small eccentricity and with numerical solutions for conical shapes with polygonal cross-sections. The solution provides the dependencies of the force on the indentation, the size and the shape of the contact area as well as the pressure distribution in the contact area. The approach is illustrated by linear (conical) and quadratic profiles with arbitrary cross-sections as well as for "separable" shapes, which can be represented as a product of a power-law function of the radius with an arbitrary exponent and an arbitrary function of the polar angle. A generalization of the Method of Dimensionality Reduction to non-axisymmetric profiles is formulated.

Keywords: contact problem; non-axisymmetric indenter; extremal principle; generalized MDR

Citation: Popov, V.L. An Approximate Solution for the Contact Problem of Profiles Slightly Deviating from Axial Symmetry. *Symmetry* **2022**, *14*, 390. https://doi.org/10.3390/sym14020390

Academic Editor: Danny Arrigo

Received: 29 January 2022
Accepted: 11 February 2022
Published: 15 February 2022
Corrected: 11 October 2022

Publisher's Note: MDPI stays neutral with regard to jurisdictional claims in published maps and institutional affiliations.

1. Introduction

In 1882, Heinrich Hertz solved the problem of elastic contact of parabolic bodies [1]. A solution for the contact of an arbitrary body of revolution with a compact contact area was found much later, in 1941–1942 by Föppl and Schubert [2–4]. An attempt to overcome the restriction of axial symmetry was undertaken in 1990 by Barber and Billings [5]. Their approach is based on using Betti's reciprocity theorem as suggested by Shield in 1967 [6] and the extremal principle found by Barber [7]. However, in [5] Barber and Billings merely illustrated their method on examples of contacts with "linear profiles" (pyramids with polygonal cross-sections) since an analytical solution is possible only for this case. In the present paper, we apply the extremal principle of Barber to another case where a largely analytical treatment is possible: To contacts of profiles that are not axially symmetrical but deviate only slightly from the axial symmetry.

2. Barber's Extremal Principle

In [6], Shield used Betti's reciprocity theorem to show that the normal force $F_N(A)$ appearing due to indentation of the profile $f(x,y)$ to a depth d (while A is the contact area in this state) is given by the equation

$$F_N(A) = \frac{1}{d^*} \iint_A p^*(x,y)(d - f(x,y))\mathrm{d}x\mathrm{d}y \tag{1}$$

where in the pressure distribution, $p^*(x,y)$ is that under a flat punch with the cross-sectional shape A which is indented by d^*. Note that the indentation d^* is an auxiliary parameter used solely for determining the pressure $p^*(x, y)$; it has nothing to do with the true indentation depth d. Of course, the integral (1) does not depend on d^* as pressure $p^*(x, y)$ under an arbitrary flat punch is proportional to d^*. In [7], Barber has shown that the area A fulfilling the usual contact conditions (the pressure inside the contact area is positive and there is no interpenetration, outside the contact, the pressure is zero and the distance between surfaces is positive), corresponds to the maximum force at a given indentation depth. (For axisymmetric profiles, this extremal principle has already been used by Shield [6]).

To be able to look constructively for a solution, Barber and Billings propose to use Fabrikant's approximation [8] for the pressure distribution under a flat-ended punch. Fabrikant's hypothesis is that the stress distribution is given in a good approximation by the equation

$$p = \frac{2E^*d}{L} \frac{a(\varphi)}{\sqrt{a(\varphi)^2 - r^2}}$$ (2)

where $a(\varphi)$ is the equation for the contact boundary in polar coordinates and

$$L = \int_0^{2\pi} a(\varphi)\mathrm{d}\varphi$$ (3)

The motivation for the ansatz (2) is straightforward. It is known to be exact for elliptical punches with arbitrary eccentricity, and it provides the correct kind of asymptotic behavior in the vicinity of the boundary, which must be fulfilled for a flat-ended punch of arbitrary shape [9].

The origin for the polar coordinate system has to be taken as the centroid of the area A provided there are no tilting moments around the origin (which we will assume in this paper).

With Fabrikant's pressure (2), Equation (1) becomes

$$F_N(A) = \frac{2E^*}{L} \int_0^{2\pi} \int_0^{a(\varphi)} \frac{a(\varphi)(d - f(r, \varphi))r\mathrm{d}r\mathrm{d}\varphi}{\sqrt{a(\varphi)^2 - r^2}}$$ (4)

Introducing definitions

$$g_\varphi(a) = a \int_0^a \frac{f'(r, \varphi)}{\sqrt{a^2 - r^2}}\mathrm{d}r, \quad \frac{\mathrm{d}G_\varphi(a)}{\mathrm{d}a} = g_\varphi(a),$$ (5)

the force (4) can be rewritten as

$$F_N(A) = \frac{2E^*}{L} \int_0^{2\pi} \left(d \cdot a(\varphi)^2 - a(\varphi)G_\varphi(a(\varphi)) \right)\mathrm{d}\varphi$$ (6)

3. Finding the Force Maximum

3.1. Axially Symmetric Profiles

In the case of an axially symmetric profile, $a(\varphi) = a_0$, $L = 2\pi a_0$, the force equation simplifies to

$$F_N(a_0) = 2E^*(d \cdot a_0 - G(a_0))$$ (7)

The condition for its maximum reads $\mathrm{d}F_N(a_0)/\mathrm{d}a_0 = 2E^*(d - g(a_0)) = 0$ or

$$d = g(a_0)$$ (8)

and the normal force is equal to

$$F_N(a_0) = E^* \int_{-a_0}^{a_0} (d_0 - g(a)) da \tag{9}$$

The pressure distribution can be calculated by integrating contributions (2) during indentation from the first touch up to the given indentation depth, as described in detail in [10] (Appendix B),

$$p(r) = \frac{E^*}{\pi} \int_{r}^{a} \frac{1}{\sqrt{\tilde{a}^2 - r^2}} \frac{dg(\tilde{a})}{d\tilde{a}} d\tilde{a} \tag{10}$$

Equations (8)–(10) are the well-known equations of the Method of Dimensionality Reduction (MDR) [11].

3.2. General Profiles

In the general case of non-axisymmetric profiles, let us first search for the maximum of functional (6) at a constant value of L. This can be done, according to Lagrange, by searching for an unconditional extremum of the functional

$$\frac{2E^*}{L} \int_{0}^{2\pi} \left(d \cdot a(\varphi)^2 - a(\varphi) G_\varphi(a(\varphi)) \right) d\varphi - D \left(\int_{0}^{2\pi} a(\varphi) d\varphi - L \right) \tag{11}$$

where D is a Lagrange multiplier. At the maximum of this functional, the first variation must vanish identically

$$\frac{2E^*}{L} \left(2d \cdot a(\varphi) - G_\varphi(a(\varphi)) - a(\varphi) g_\varphi(a(\varphi)) \right) = D \tag{12}$$

This equation determines implicitly the contact boundary $a(\varphi)$, while D is connected with L through the condition (3). After substitution of $a(\varphi)$ into (6), the force will remain a function of the still undefined L, which should finally be found as a value giving the maximum to the force. All stated operations, beginning with the solution of Equation (12) with respect to $a(\varphi)$, can be carried out analytically without approximations only in two cases: Either the above case of axisymmetric profiles (leading to the MDR) or in the case of linear (conical) profiles. The latter has already been done in the paper [5] for polygonal cross-sections. To be able to constructively realize the above program for more general shapes, let us consider profiles that only slightly deviate from axisymmetric ones. We will show below that this is the third case where a largely analytical treatment is possible.

3.3. Profiles Slightly Deviating from Axisymmetric Ones

Let us consider a profile that is not axially symmetrical but deviates only slightly from axial symmetry:

$$f(r, \varphi) = f_0(r) + \delta f(r, \varphi) \tag{13}$$

where $\delta f(r, \varphi)$ is a small deviation. We define the position of the origin of coordinates in such a way that

$$f(0, \varphi) = f_0(0) = \delta f(0, \varphi) = 0 \tag{14}$$

The functions $g_\varphi(a)$ and $G_\varphi(a)$ will correspondingly consist of two parts

$$G_\varphi(a) = G_0(a) + \delta G_\varphi(a), \ g_\varphi(a) = g_0(a) + \delta g_\varphi(a), \ g_\varphi{}'(a) = g_0{}'(a) + \delta g_\varphi{}'(a) \tag{15}$$

with

$$g_0(a) = a \int_{0}^{a} \frac{f_0{}'(r)}{\sqrt{a^2 - r^2}} dr, \ \delta g_\varphi(a) = a \int_{0}^{a} \frac{\delta f_0{}'(r, \varphi)}{\sqrt{a^2 - r^2}} dr \tag{16}$$

and correspondingly

$$G_0{}'(a) = g_0(a), \ \delta G'{}_\varphi(a) = \delta g_\varphi(a) \tag{17}$$

The contact boundary will also be almost a circle with a small perturbation

$$a(\varphi) = a_0 + \delta a(\varphi) \tag{18}$$

Substituting (18) into (12), expanding all functions with respect to small deviation $\delta a(\varphi)$ and neglecting all terms of the second or higher orders, we get for the deviation of the contact contour from a circle in the first approximation

$$\delta a(\varphi) = \frac{\frac{D\pi a_0}{E^*} - 2da_0 + G_\varphi(a_0) + a_0 g_\varphi(a_0)}{2d - 2g_\varphi(a_0) - a_0 g_\varphi{}'(a_0)} \tag{19}$$

With (15), we obtain

$$\begin{aligned}
\delta a(\varphi) &[2d - 2g_0(a_0) - a_0 g_0{}'(a_0)]^2 \\
&= \left[\frac{D\pi a_0}{E^*} - 2da_0 + G_0(a_0) + a_0 g_0(a_0)\right][2d - 2g_0(a_0) - a_0 g_0{}'(a_0)] \\
&+ [2d - 2g_0(a_0) - a_0 g_0{}'(a_0)][\delta G_\varphi(a_0) + a_0 \delta g_\varphi(a_0)] \\
&+ \left[\frac{D\pi a_0}{E^*} - 2da_0 + G_0(a_0) + a_0 g_0(a_0)\right][2\delta g_\varphi(a_0) - a_0 \delta g_\varphi{}'(a_0)]
\end{aligned} \tag{20}$$

To determine the Lagrange multiplier D we require

$$\int_0^{2\pi} \delta a(\varphi)\mathrm{d}\varphi = 0 \tag{21}$$

Note that the separation (13) into an axisymmetric part and perturbation is not unique. We can use this freedom to essentially simplify the following relations. Let us define the deviation in such a way that

$$\int_0^{2\pi} \delta g_\varphi(a_0)\mathrm{d}\varphi = 0 \tag{22}$$

for all a_0. This will automatically mean that

$$\int_0^{2\pi} \delta G_\varphi(a_0)\mathrm{d}\varphi = 0 \text{ and } \int_0^{2\pi} \delta g_\varphi{}'(a_0)\mathrm{d}\varphi = 0 \tag{23}$$

and also guarantee that

$$\int_0^{2\pi} [\delta G_\varphi(a_0) + a_0 \delta g_\varphi(a_0)]\mathrm{d}\varphi = 0 \tag{24}$$

and thus, that integral over φ of the term in the third line in (20) vanishes. If we now chose the constant D such that $\left[\frac{D\pi a_0}{E^*} - 2da_0 + G_0(a_0) + a_0 g_0(a_0)\right] = 0$, then the integrals over φ of the terms in the second and fourth lines of (20) vanish identically and the condition (21) is fulfilled. We then obtain from (20)

$$\delta a(\varphi) = \frac{[\delta G_\varphi(a_0) + a_0 \delta g_\varphi(a_0)]}{[2d - 2g_0(a_0) - a_0 g_0{}'(a_0)]} \tag{25}$$

The force (6) can now be written as

$$F_N(A) = \frac{E^*}{\pi a_0} \int\limits_0^{2\pi} \left(d \cdot a(\varphi)^2 - a(\varphi) G_\varphi(a(\varphi)) \right) d\varphi$$

$$= \frac{E^*}{\pi a_0} \int\limits_0^{2\pi} \left(d \cdot a_0^2 - a_0 G_0(a_0) \right) d\varphi + \frac{E^*}{\pi a_0} \int\limits_0^{2\pi} [2d \cdot a_0 - G_0(a_0) - a_0 g_0(a_0)] \delta a(\varphi) d\varphi - \frac{E^*}{\pi} \int\limits_0^{2\pi} \delta G_\varphi(a_0) d\varphi \tag{26}$$

The second and the third term are equal to zero due to relations (22) and (23) so that only the first term remains. Its maximization leads to the usual MDR result [11]

$$d = g_0(a_0) \tag{27}$$

This means that Equation (25) can finally be rewritten as follows:

$$\delta a(\varphi) = -\frac{\delta G_\varphi(a_0) + a_0 \delta g_\varphi(a_0)}{a_0 g_0'(a_0)} \tag{28}$$

This equation provides the explicit solution for the deviation of the contact boundary from the circle with radius a_0. The normal force is given by the Equation (9)

$$F_N(a_0) - 2E^* \int\limits_0^{a_0} (d - g_0(a)) da \tag{29}$$

3.1. Pressure Distribution for Profiles Slightly Deviating from Axisymmetric Ones

Let us take a close look at the process of indentation from the first touch to the final indentation depth d and denote the current values of the force, the indentation depth and the effective contact radius respectively by $\tilde{F}_N$, $\tilde{d}$ and $\tilde{a}_0$. The entire process consists of changing the indentation depth from $\tilde{d} = 0$ to $\tilde{d} = d$, whereby the contact radius changes from $\tilde{a} = 0$ to $\tilde{a}_0 = a_0$ and the contact force from $\tilde{F}_N = 0$ to $\tilde{F}_N = F_N$. An infinitesimal indentation by $d\tilde{d}$ of the area, which is given by the contour equation $r = a(\varphi)$, produces the following contribution to the pressure distribution

$$dp = \frac{E^*}{\pi a_0} \frac{a(\varphi)}{\sqrt{a(\varphi)^2 - r^2}} d\tilde{d} \text{ for } r < a(\varphi) \tag{30}$$

The pressure distribution at the end of the indentation process is equal to the sum of the incremental pressure distributions:

$$p(r, \varphi) = \frac{E^*}{\pi} \int\limits_{d(r)}^{d} \frac{1}{\tilde{a}_0} \frac{\tilde{a}(\varphi)}{\sqrt{\tilde{a}(\varphi)^2 - r^2}} d\tilde{d} = \frac{E^*}{\pi} \int\limits_{r}^{a(\varphi)} \frac{1}{\sqrt{\tilde{a}(\varphi)^2 - r^2}} \frac{\tilde{a}(\varphi)}{\tilde{a}_0} \frac{dg_0(\tilde{a}_0)}{d\tilde{a}(\varphi)} d\tilde{a}(\varphi) \tag{31}$$

where $\tilde{a}_0$ must be considered here as a function of $\tilde{a}(\varphi)$.

4. Examples

4.1. Contact with Parabolic Profiles with Arbitrary Cross-Section

Consider a profile

$$z = f(r, \varphi) = r^2 \psi(\varphi) \tag{32}$$

Thus

$$f_0(r) = r^2 \overline{\psi} \text{ and } \delta f(r, \varphi) = r^2 (\psi(\varphi) - \overline{\psi}) \tag{33}$$

where $\overline{\psi}$ is the averaged value of $\psi(\varphi)$ over all angles:

$$\overline{\psi} = \frac{1}{2\pi} \int\limits_0^{2\pi} \psi(\varphi) d\varphi \tag{34}$$

The corresponding angle-dependent "MDR profile" is

$$g_\varphi(a) = 2a^2 \psi(\varphi) \tag{35}$$

For all necessary auxiliary functions, we get

$$g_0(a) = 2\overline{\psi} a^2, \ \delta g_\varphi(a) = 2a^2 \left(\psi(\varphi) - \overline{\psi}\right), \ g_0'(a) = 4\overline{\psi} a, \tag{36}$$

$$G_0(a) = \frac{2\overline{\psi} a^3}{3}, \ \delta G_\varphi(a) = \frac{2a^3}{3}\left(\psi(\varphi) - \overline{\psi}\right). \tag{37}$$

The effective contact radius is given by Equation (27):

$$a_0 = \sqrt{\frac{d}{2\overline{\psi}}} \tag{38}$$

and the contact area is given by the relation (18)

$$a(\varphi) = a_0 \left(\frac{5}{3} - \frac{2}{3}\frac{\psi(\varphi)}{\overline{\psi}}\right) \tag{39}$$

To find the pressure distribution, we use Equation (31)

$$p(r,\varphi) = \frac{E^*}{\pi} \int_r^{a(\varphi)} \frac{\tilde{a}(\varphi)}{\sqrt{\tilde{a}(\varphi)^2 - r^2}} \frac{1}{\tilde{a}_0} \frac{\mathrm{d}g_0(\tilde{a}_0)}{\mathrm{d}\tilde{a}(\varphi)} \mathrm{d}\tilde{a}(\varphi) = \frac{2}{\pi}E^*\left(2d\cdot\overline{\psi}\right)^{1/2}\sqrt{1 - \left(\frac{r}{a(\varphi)}\right)^2} \tag{40}$$

4.2. Contact with a Paraboloid

As a special case of a general parabolic profile, let us consider a paraboloid

$$z = f(x,y) = \frac{x^2}{2R_1} + \frac{y^2}{2R_2} = \frac{r^2}{2}\left(\frac{\cos^2\varphi}{R_1} + \frac{\sin^2\varphi}{R_2}\right) \tag{41}$$

In this case,

$$\psi(\varphi) = \frac{1}{2}\left(\frac{\cos^2\varphi}{R_1} + \frac{\sin^2\varphi}{R_2}\right) = \frac{1}{4}\left(\frac{1}{R_1} + \frac{1}{R_2}\right) + \frac{1}{4}\left(\frac{1}{R_1} - \frac{1}{R_2}\right)\cos 2\varphi \tag{42}$$

and

$$\overline{\psi} = \frac{1}{4}\left(\frac{1}{R_1} + \frac{1}{R_2}\right) \tag{43}$$

Equations (38)–(40) take the form

$$a_0 = \sqrt{\frac{2d \cdot R_1 R_2}{R_1 + R_2}} \tag{44}$$

$$a(r,\varphi) = a_0 \left[1 - \frac{2}{3}\frac{R_2 - R_1}{R_1 + R_2}\cos 2\varphi\right] \tag{45}$$

and

$$p(r,\varphi) = \frac{2}{\pi}E^*\left(\frac{d\cdot(R_1 + R_2)}{2R_1 R_2}\right)^{1/2}\sqrt{1 - \left(\frac{r}{a(\varphi)}\right)^2} \tag{46}$$

The total normal force is equal to

$$F_N = \frac{4}{3}E^*\left(\frac{2R_1 R_2}{R_1 + R_2}\right)^{1/2} d^{3/2}\left(1 - \left(\frac{2}{3}\frac{R_2 - R_1}{R_1 + R_2}\right)^2\right) \tag{47}$$

Equation (45) describes in the approximation of small eccentricities an ellipse with half-axes

$$a = a_0\left[1 + \frac{2}{3}\frac{R_2 - R_1}{R_1 + R_2}\right], \text{ and } b = a_0\left[1 - \frac{2}{3}\frac{R_2 - R_1}{R_1 + R_2}\right] \tag{48}$$

For the eccentricity of the contact area, we get

$$e = \sqrt{1 - \frac{b^2}{a^2}} = \frac{2}{\sqrt{3}}e_g\frac{\sqrt{1 - e_g^2/2}}{1 - e_g^2/6} \tag{49}$$

where

$$e_g = \sqrt{1 - \frac{R_1}{R_2}} \tag{50}$$

is the eccentricity of cross-sections of the indenter. For small eccentricities, the ratio of eccentricities of contact area and cross-sections of the ellipsoid is constant and equal to $2/\sqrt{3}$, which is the exact asymptotic value according to the Hertzian solution [9] (p. 35). Equation (49) provides a very good approximation for the exact solution for eccentricities up to 0.7 (see Figure 1).

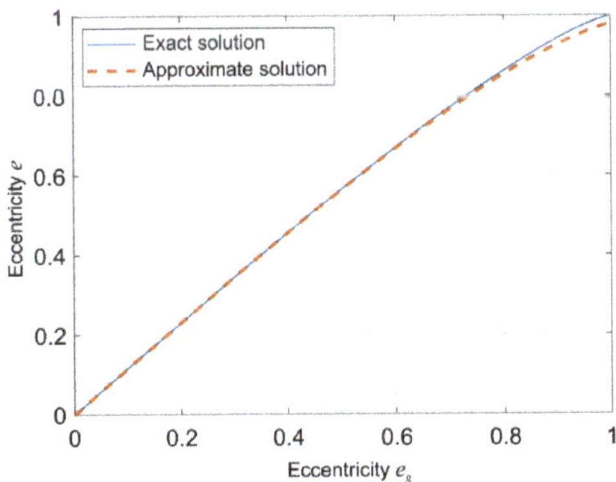

Figure 1. Eccentricity e of the contact area as a function of e_g according to Equation (49) (dashed line) and comparison with the value according to the Hertzian solution [9] (solid line).

4.3. Contact of a Conical Indenter with an Arbitrary Cross-Section

Consider a general conical profile with an "almost axisymmetric" shape

$$f(r, \varphi) = r \cdot \psi(\varphi) = f_0(r) + \delta f(r, \varphi) \tag{51}$$

Thus

$$f_0(r) = r\overline{\psi} \text{ and } \delta f(r, \varphi) = r\left(\psi(\varphi) - \overline{\psi}\right) \tag{52}$$

where $\overline{\psi}$ is the averaged value of $\psi(\varphi)$ over all angles according to (34). For all necessary auxiliary functions, we get

$$g_0(a) = \frac{\pi}{2}a\overline{\psi}, \; \delta g_\psi(a) - \frac{\pi}{2}a\left(\psi(\varphi) - \overline{\psi}\right), \; g_0'(a) = \frac{\pi}{2}\overline{\psi} \tag{53}$$

$$G_0(a) = \frac{\pi}{4}a^2\overline{\psi}, \; \delta G_\varphi(a) = \frac{\pi}{4}a^2\left(\psi(\varphi) - c\right) \tag{54}$$

The contact area is given by the relation (18)

$$a(\varphi) = \frac{2}{\pi}\frac{d}{\overline{\psi}}\left(\frac{5}{2} - \frac{3}{2}\frac{\psi(\varphi)}{\overline{\psi}}\right) \tag{55}$$

To find the pressure distribution, we use Equation (31)

$$p(r,\varphi) = \frac{E^*}{\pi} \int_r^{a(\varphi)} \frac{1}{\sqrt{\tilde{a}(\varphi)^2 - r^2}} \frac{\tilde{a}(\varphi)}{\tilde{a}_0} \frac{\mathrm{d}g_0(\tilde{a}_0)}{\mathrm{d}\tilde{a}(\varphi)} \mathrm{d}\tilde{a}(\varphi) = \frac{E^*}{2}\overline{\psi}\cdot\ln\left(\frac{a(\varphi)}{r} + \sqrt{\left(\frac{a(\varphi)}{r}\right)^2 - 1}\right) \tag{56}$$

4.4. Contact of a Pyramid with Square Cross-Section

Let's consider a linear profile

$$f(r,\varphi) = r\tan\alpha\cos\varphi, \ -\frac{\pi}{4} < \varphi < \frac{\pi}{4} \tag{57}$$

and similar equations for the angles $\pi/4 < \varphi < 3\pi/4$, $3\pi/4 < \varphi < 5\pi/4$ and $-3\pi/4 < \varphi < -\pi/4$, which describes a pyramid with square cross-sections and the angle α between the inclined planes of the pyramid and the horizon. In this case,

$$\psi(\varphi) = \tan\alpha\cos\varphi, \ \overline{\psi} = \frac{2\sqrt{2}}{\pi}\tan\alpha \tag{58}$$

The boundary of the contact area is given by the relation (55) (see Figure 2)

$$a(\varphi) = \frac{d}{\tan\alpha}\left(\frac{5}{2\sqrt{2}} - \frac{3\pi}{8}\cos\varphi\right) \text{ for } -\frac{\pi}{4} < \varphi < \frac{\pi}{4} \text{ usw.} \tag{59}$$

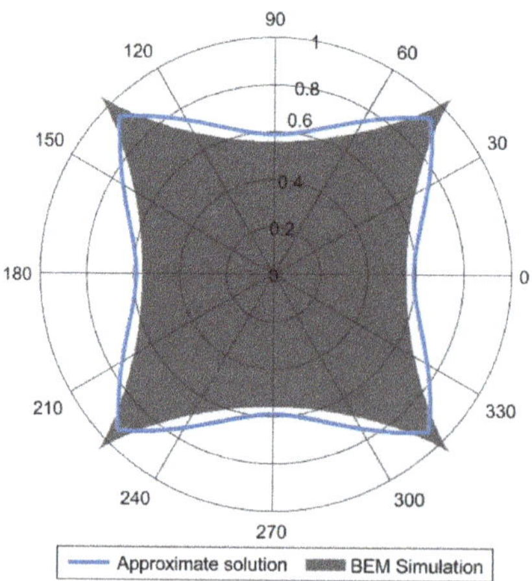

Figure 2. The normalized contact boundary line $(a(\varphi)\tan\alpha/d)$ according to Equation (59) (blue line) and the result of numerical simulation with Boundary Element Method (gray figure). Numerical data have been obtained by Qiang Li [12] with the method described in [13].

For the normal force, we get

$$F_N = \frac{2}{\pi}E^*\frac{d^2}{\overline{\psi}} = \frac{E^*d^2}{\sqrt{2}\tan\alpha} \tag{60}$$

Numerical simulation gives the solution $F_N = 0.7395E^*d^2/\tan\alpha$ which is 4.6% larger than the above analytical result.

For the pressure distribution in the contact area, we obtain, according to (56),

$$p(r,\varphi) = \frac{E^*}{2}\overline{\psi}\cdot\ln\left(\frac{a(\varphi)}{r} + \sqrt{\left(\frac{a(\varphi)}{r}\right)^2 - 1}\right) \tag{61}$$

4.5. Contact of Power-Law Shapes

Consider a profile having the form

$$f(r, \varphi) = r^n \cdot \psi(\varphi) \tag{62}$$

which means that all vertical cross-sections are self similar differing only by the scaling factor $\psi(\varphi)$. The decomposition into a rotationally symmetric part and deviation is as follows

$$f_0(r) = r^n \cdot \overline{\psi}, \ \delta f(r, \varphi) = r^n \cdot \left(\psi(\varphi) - \overline{\psi}\right) \tag{63}$$

where $\overline{\psi}$ is the average value of $\psi(\varphi)$, Equation (34).

For all necessary auxiliary functions, we get

$$g_\varphi(a) = \psi(\varphi) \cdot \gamma(a), \ G_\varphi(a) = \psi(\varphi) \cdot (a), \ '(a) = \gamma(a), \ g_0(a) = \overline{\psi} \cdot \gamma(a) \tag{64}$$

with

$$\gamma(a) = a \int_0^a \frac{n r^{n-1}}{\sqrt{a^2 - r^2}} dr = \kappa_n a^n, \ \kappa_n = \int_0^1 \frac{\zeta^{n-1} d\zeta}{\sqrt{1 - \zeta^2}} = \frac{\sqrt{\pi}}{2} \frac{n\Gamma\left(\frac{n}{2}\right)}{\Gamma\left(\frac{n}{2} + \frac{1}{2}\right)} \tag{65}$$

while $\Gamma(\ldots)$ is the gamma function.

For the deviations, we have

$$\delta g_\varphi(a) = \kappa_n a^n \left(\psi(\varphi) - \overline{\psi}\right), \ \delta G_\varphi(a) = \kappa_n \frac{a^{n+1}}{n+1} \left(\psi(\varphi) - \overline{\psi}\right) \tag{66}$$

The boundary of the contact area is given by the relation (28)

$$a(\varphi) = a_0 \left(1 + \frac{n+2}{n(n+1)}\left(1 - \frac{\psi(\varphi)}{\overline{\psi}}\right)\right) \tag{67}$$

where a_0 is determined by the condition $\overline{\psi}\kappa_n a_0{}^n = d$:

$$a_0 = \left(\frac{d}{\kappa_n \overline{\psi}}\right)^{1/n} \tag{68}$$

The normal force is equal to

$$F_N(a_0) = 2E^* \int_0^{a_0} \left(d - \overline{\psi}\kappa_n a^n\right) da = 2E^* d a_0 \frac{n}{n+1} = \frac{2n}{n+1} E^* d^{\frac{n+1}{n}} \left(\kappa_n \overline{\psi}\right)^{-1/n} \tag{69}$$

The average pressure is equal to

$$\overline{p} = \frac{F_N(a_0)}{\pi a_0^2} = \frac{2n}{n+1} \frac{E^* d^{\frac{n-1}{n}} \left(\kappa_n \overline{\psi}\right)^{1/n}}{\pi} \tag{70}$$

For the pressure distribution in the contact area, normalized by the average pressure, we obtain, according to (31),

$$p(r, \varphi) = \frac{(n+1)}{2} \int_{\overline{r}}^1 \frac{\zeta^{n-1}}{\sqrt{\zeta^2 - \overline{r}^2}} d\zeta \tag{71}$$

with $\overline{r} = r/a(\varphi)$. It is again the same result as that for an axisymmetric contact ([10], p. 78) with the substitution $r/a_0 \to r/a(\varphi)$.

5. Extended Method of Dimensionality Reduction (MDR) for Slightly Non-Axisymmetric Contacts

The above-sketched solution procedure can be considered as an extension of the Method of Dimensionality Reduction [11] to non-axisymmetric profiles. Let us summarize the above findings in the form of procedure that has to be applied for the solution of such problems.

We consider an "arbitrary" profile $z = f(r, \varphi)$ underlying the following restrictions:

(1) the profile deviates only weakly from an axisymmetric one;
(2) during the normal indentation, only the normal contact force appears (no tilting forces or moments).

Under these conditions, the following procedure can be applied.

I. In the first step, an "equivalent axisymmetric profile" is determined. In the first-order approximation, it was shown to be just the profile averaged over the angles:

$$f_0(r) = \frac{1}{2\pi} \int_0^{2\pi} f(r, \varphi) \mathrm{d}\varphi \tag{72}$$

II. The profile can now be decomposed into an axisymmetric part and the small deviation

$$f(r, \varphi) = f_0(r) + \delta f(r, \varphi) \tag{73}$$

with

$$\delta f(r, \varphi) = f(r, \varphi) - f_0(r) \tag{74}$$

by definition.

III. With the equivalent axisymmetric profile (72), the usual MDR solution procedure is applied. In particular, the MDR-transformed profile is determined as

$$g_0(a) = a \int_0^a \frac{f_0'(r)}{\sqrt{a^2 - r^2}} \mathrm{d}r \tag{75}$$

IV. The effective radius a_0 is determined by the condition

$$d = g_0(a_0) \tag{76}$$

V. The normal force is given by the equation

$$F_N = 2E^* \int_0^{a_0} (d - g_0(x)) \mathrm{d}x \tag{77}$$

VI. The effective pressure distribution in the contact area is given by the usual MDR equation [14] (p. 9)

$$p_0(r/a_0) = -\frac{1}{\pi} \int_r^\infty \frac{q'(x)}{\sqrt{x^2 - r^2}} \mathrm{d}x \tag{78}$$

with

$$q(x) = E^* \begin{cases} d - g(x), |x| < a_0 \\ 0, |x| > a_0 \end{cases} \tag{79}$$

VII. The true non-axisymmetric contact area is given by the equation

$$a(\varphi) = a_0 + \delta a(\varphi) \tag{80}$$

where $\delta a(\varphi)$ is determined by (28).

VIII. Finally, the true pressure distribution is given by

$$p(r, \varphi) = p_0(r/a(\varphi)) \tag{81}$$

The last equation can be considered as a generalization of Fabrikant's ansatz for flat-ended punches. Fabrikant states that the pressure distribution under a non-axisymmetric punch is equal to that under an axisymmetric punch but "rescaled" to the true shape of the contact area. Similarly, Equation (81) states that the contact pressure is equal to that under an "equivalent axisymmetric indenter" but rescaled to the true contact area. In the present paper, we have shown that Equation (81) is the exact first-order approximation for the power-law profiles (with arbitrary cross-section)—independently of the exponent of the power law. Even while it was not proven in a general case, the independence of the exponent gives hope that it could be a good approximation for arbitrarily shaped profiles. Testing of this hypothesis (e.g., through comparison with a numerical solution with BEM) is an important task of further studies.

6. Discussion

Using the Barber's extremal principle, we derived explicit analytical relations for all essential contact properties of an indenter of "arbitrary" shape (under restriction that it should be close to an axisymmetric one). The solution provides the dependency of the normal force on the indentation depth, the size and the shape of the contact area, and the pressure distribution in the contact area. The derivation has been carried out under two assumptions (which are the main sources of deviation from the exact solution): (1) Fabrikant's approximation for the stress of a flat-ended punch with arbitrary cross-section and (2) assumption of a small deviation of the indenter shape from an axial one. However, an acceptable accuracy is obtained even with relatively large deviations from axial symmetry, e.g., the eccentricity of the contact area for a paraboloid is given accurately up to an eccentricity of approximately 0.7. The deviation of normal force in a contact of a pyramid indenter with square cross-section is of about 4.6%. The whole calculation resembles the Method of Dimensionality Reduction very much and can be considered as its generalization for non-axially symmetric contacts. A central approximation used in this paper is Fabrikant's approximation for the pressure under a flat punch. This approximation could be further improved by using higher-order corrections obtained by Golikova and Mossakovskii for the pressure distribution under plane stamps of nearly circular cross-section [15].

Funding: This work has been conducted under partial financial support from the German Research Society (DFG), Project PO 810/66-1. It was partially supported by the Tomsk State University Development Programme («Priority-2030»).

Institutional Review Board Statement: Not applicable.

Informed Consent Statement: Not applicable.

Data Availability Statement: Not applicable.

Acknowledgments: The author thanks E. Willert for discussions and Q. Li for providing numerical data for indentation of a pyramid. We acknowledge support by the German Research Foundation and the Open Access Publication Fund of TU Berlin.

Conflicts of Interest: The author declares no conflict of interest.

References

1. Hertz, H. Über die Berührung fester elastischer Körper. *J. Für Die Reine Und Angew. Math.* **1882**, *92*, 156–171. [CrossRef]
2. Popova, E.; Popov, V.L. Ludwig Föppl and Gerhard Schubert: Unknown classics of contact mechanics. *ZAMM-J. Appl. Math. Mech./Z. Für Angew. Math. Und Mech.* **2020**, *100*, e202000203. [CrossRef]
3. Föppl, L. Elastische Beanspruchung des Erdbodens unter Fundamenten. *Forsch. Auf Dem Geb. Des Ing. A* **1941**, *12*, 31–39. [CrossRef]
4. Schubert, G. Zur Frage der Druckverteilung unter elastisch gelagerten Tragwerken. *Ing.-Arch.* **1942**, *13*, 132–147. [CrossRef]

5. Barber, J.R.; Billings, D.A. An approximative solution for the contact area and elastic compliance of a smooth punch of arbitrary shape. *Int. J. Mech. Sci.* **1990**, *32*, 991–997. [CrossRef]
6. Shield, R.T. Load-displacement relations for elastic bodies. *Z. Für Angew. Math. Und Phys. ZAMP* **1967**, *18*, 682–693. [CrossRef]
7. Barber, J.R. Determining the contact area in elastic indentation problems. *J. Strain Anal.* **1974**, *9*, 230–232. [CrossRef]
8. Fabrikant, V.I. Flat punch of arbitrary shape on an elastic half-space. *Int. J. Eng. Sci.* **1986**, *24*, 1731–1740. [CrossRef]
9. Barber, J.R. *Contact Mechanics*; Springer: Cham, Switzerland, 2018; p. 585. [CrossRef]
10. Popov, V.L. *Contact Mechanics and Friction. Physical Principles and Applications*, 2nd ed.; Springer: Berlin/Heidelberg, Germany, 2017. [CrossRef]
11. Popov, V.L.; Heß, M. *Method of Dimensionality Reduction in Contact Mechanics and Friction*; Springer: Berlin/Heidelberg, Germany, 2015; 265p. [CrossRef]
12. Li, Q.; Technische Universität, Berlin, Germany. Personal Communication, 2022.
13. Pohrt, R.; Li, Q. Complete boundary element formulation for normal and tangential contact problems. *Phys. Mesomech.* **2014**, *17*, 334–340. [CrossRef]
14. Popov, V.L.; Heß, M.; Willert, E. *Handbook of Contact Mechanics. Exact Solutions of Axisymmetric Contact Problems*; Springer: Berlin/Heidelberg, Germany, 2019. [CrossRef]
15. Golikova, S.S.; Mossakovskii, V.I. Pressure of a plane stamp of nearly circular cross section on an elastic half-space. *PMM J. Appl. Math. Mech.* **1976**, *40*, 1085–1088. [CrossRef]

 *symmetry*

Article

A Comparison of General Solutions to the Non-Axisymmetric Frictionless Contact Problem with a Circular Area of Contact: When the Symmetry Does Not Matter

Ivan Argatov

Institut für Mechanik, Technische Universität Berlin, 10623 Berlin, Germany; ivan.argatov@campus.tu-berlin.de

Abstract: The non-axisymmetric problem of frictionless contact between an isotropic elastic half-space and a cylindrical punch with an arbitrarily shaped base is considered. The contact problem is formulated as a two-dimensional Fredholm integral equation of the first type in a fixed circular domain with the right-hand side being representable in the form of a Fourier series. A number of general solutions of the contact problem, which were published in the literature, are discussed. Based on the Galin–Mossakovskii general solution, new formulas are derived for the particular value of the contact pressure at the contact center and the contact stress-intensity factor at the contour of the contact area. Since the named general solution does not employ the operation of differentiation of a double integral with respect to the coordinates that enter it as parameters, the form of the general solution derived by Mossakovskii as a generalization of Galin's solution for the special case, when the contact pressure beneath the indenter is bounded, is recommended for use as the most simple closed-form general solution of the non-axisymmetric Boussinesq contact problem.

Keywords: contact problem; non-axisymmetric; circular contact; frictionless contact; general solution; closed-form solution; series solution; cylindrical punch; contact stress-intensity factor

Citation: Argatov, I. A Comparison of General Solutions to the Non-Axisymmetric Frictionless Contact Problem with a Circular Area of Contact: When the Symmetry Does Not Matter. *Symmetry* 2022, 14, 1083. https://doi.org/10.3390/sym14061083

Academic Editor: Jan Awrejcewicz

Received: 16 April 2022
Accepted: 17 May 2022
Published: 25 May 2022

Publisher's Note: MDPI stays neutral with regard to jurisdictional claims in published maps and institutional affiliations.

1. Introduction

The contact mechanics dates back to Hertz (1882), who developed the theory of unilateral frictionless local contact for two elastic bodies, which in the unloaded state are shaped as elliptic paraboloids in the vicinity of the point of initial contact, and Boussinesq (1885), who solved the problem of contact between an elastic half-space and a frictionless flat-ended cylindrical punch. Since then, particular progress has been made with regard to solving the axisymmetric frictionless problems with a circular area of contact.

While in the literature, the general solution of the axisymmetric problem is usually associated with Sneddon's paper [1] of 1965, different authors give the priority to other studies. In particular, in his comprehensive review, Borodich [2] highlighted the contribution made by Galin (1946); Barber [3] in his book referred to the general solution as the Green and Collins solution; in their historical note [4] (see also [5]), Popova and Popov, acknowledging the contributions made by Galin and Sneddon, put under a spotlight the original paper [6] written in German by Schubert in 1942. However, for the sake of historical truth, it should be underlined that (to the best of the author's knowledge) the priority of solving the axisymmetric frictionless contact problem with a circular contact area belongs to Leonov's paper [7] published in Russian in 1939. As it was shown by Argatov and Dmitriev [8], other forms of the general solution follow from Leonov's results by the simple change of integration variables. As a compromise, Argatov and Mishuris [9] suggested to call the general solution of the axisymmetric frictionless contact problem the Galin–Sneddon solution.

The general solution of the non-axisymmetric contact problem for a cylindrical indenter is of great importance in developing the contact stiffness indentation tomography technique [10,11]. Another example of the application of the general solution is given by the problem of adhesive contact under non-symmetric perturbation of the contact geometry [12]. Generally speaking, the general solutions collected below will be useful in solving the frictionless contact problems with a circular area of contact (e.g., with applications in geotechnics [13,14]), when the symmetry of the contact geometry does not matter.

Whereas axisymmetric contact problems are considered in many publications, including textbooks [3,8] and handbook [15], the situation with the non-axisymmetric contact problem with a circular area of contact is not so equivocal, even in spite of the fact that this problem is a direct generalization of the Boussinesq problem for a cylindrical punch with a non-flat base. This paper aims to bridge this gap by comparing different general solutions published in the literature.

The main motivation for writing this reviewer paper was to identify in a sense the simplest closed form of the general solution. Another quite utilitarian motivation was to collect in one compendium the practically useful results, some of which are not readily accessible. Herein, we compare only the solutions collected from the literature, and the discussion of the methods of their derivations falls outside the scope of the present study. The recent paper [16] on solving Keer's indentation problem for a cylindrical indenter with the face in a wedge form can be regarded as a case study for the use of the general solutions.

2. General Solutions of the Frictionless Non-Axisymmetric Contact Problem

2.1. The Boussinesq Contact Problem Formulation

We consider the so-called Boussinesq contact problem for an isotropic elastic half-space (see Figure 1), which is indented by a frictionless cylindrical punch of radius a with a non-flat base described by a continuous shape function, $\Phi(r, \varphi)$. For the sake of simplicity we assume that the center of cylindrical coordinates (r, φ, z) is taken at the center of the circular area of contact, and the elastic semi-infinite body occupies the half-space $z \geq 0$.

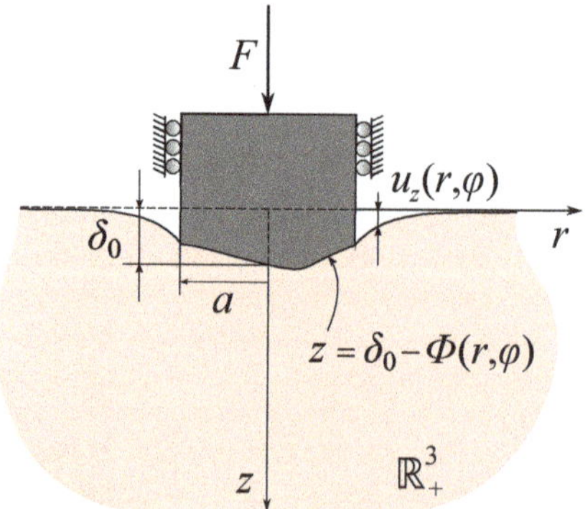

Figure 1. A schematic of the non-axisymmetric Boussinesq contact problem with a circular area of contact.

Let us assume that the shape function $\Phi(r, \varphi)$ admits the Fourier series representation

$$\Phi(r, \varphi) = \Phi_0(r) + \sum_{n=1}^{\infty} \Phi_n^s(r) \sin n\varphi + \Phi_n^c(r) \cos n\varphi, \tag{1}$$

where

$$\Phi_0(r) = \frac{1}{2\pi} \int_0^{2\pi} \Phi(r, \varphi)\, d\varphi, \quad \begin{Bmatrix} \Phi_n^s(r) \\ \Phi_n^c(r) \end{Bmatrix} = \frac{1}{\pi} \int_0^{2\pi} \Phi(r, \varphi) \begin{Bmatrix} \sin n\varphi \\ \cos n\varphi \end{Bmatrix} d\varphi. \tag{2}$$

Since the function $\Phi(r, \varphi)$ is assumed to be continuous, it can be shown that $\Phi_n^s(0) = \Phi_n^c(0) = 0$. In addition, we put

$$\Phi_0(0) = 0. \tag{3}$$

Further, let $p(r, \varphi)$ denote the contact pressure exerted by the punch under the action of an external load, F. Then, the condition of static equilibrium implies that

$$F = \int_0^{2\pi} \int_0^a p(r, \varphi) r \, dr \, d\varphi. \tag{4}$$

According to the Boussinesq solution of the problem of normal loading of an elastic half-space, the contact pressure $p(r, \varphi)$ produces the following normal surface displacement field:

$$u_z(r, \varphi) = \frac{1}{\pi E^*} \int_0^{2\pi} \int_0^a \frac{p(\rho, \varphi)\rho \, d\rho \, d\psi}{\sqrt{r^2 + \rho^2 - 2r\rho \cos(\varphi - \psi)}}. \tag{5}$$

Here, $E^* = E/(1 - v^2)$ is the reduced elastic modulus, E and v are Young's modulus and Poisson's ratio of the elastic semi-infinite body, (r, φ) are coordinates of the point of observation, and (ρ, φ) are coordinates of the point of integration.

Inside the contact area, the normal surface displacements are determined by the shape of the punch, which under the applied load receives some vertical (normal) displacement, δ_0, such that

$$u_z(r, \varphi) = \delta_0 - \Phi(r, \varphi), \quad r \in [0, a], \ \varphi \in [0, 2\pi). \tag{6}$$

Thus, from (5) and (6), it follows that

$$\frac{1}{\pi E^*} \int_0^{2\pi} \int_0^a \frac{p(\rho, \varphi)\rho \, d\rho \, d\psi}{\sqrt{r^2 + \rho^2 - 2r\rho \cos(\varphi - \psi)}} = \delta_0 - \Phi(r, \varphi), \tag{7}$$

where $r \in [0, a]$ and $\varphi \in [0, 2\pi)$.

The integral Equation (7) is called the governing integral equation of the Boussinesq contact problem with the circular contact area of radius a. It is to emphasize that in contrast to the Hertz contact problem, where the contact area is determined from the condition of vanishing of the contact pressure on the contour of the contact area, in the Boussinesq contact problem, the contact area is assumed to be a priori fixed. Correspondingly, the contact pressure density $p(r, \varphi)$, which solves Equation (7), may possess a singularity at the contact contour, as the point of observation (r, φ) approaches the contact contour, when $r \to a$.

To simplify formulas, we introduce the auxiliary notation

$$f(r, \varphi) = \delta_0 - \Phi(r, \varphi) \tag{8}$$

$$= f_0(r) + \sum_{n=1}^{\infty} f_n^s(r) \sin n\varphi + f_n^c(r) \cos n\varphi, \tag{9}$$

where, in view of (1), we have

$$f_0(r) = \delta_0 - \Phi_0(r), \quad f_n^s(r) = -\Phi_n^s(r), \quad f_n^c(r) = -\Phi_n^c(r). \tag{10}$$

We note that a rigid body, by means of which external loads are transferred to the surface of an elastic body, is usually called an indenter. Here we prefer the term 'punch', as the term 'indenter' (in many cases) assumes unilateral contact, when only positive contact pressures are allowed inside the contact area.

Finally, we note that, in view of (3), the parameter δ_0 has the exact meaning of the normal surface displacement at the center of the contact area. If $\Phi(r, \varphi) \geq 0$ for any $r \in [0, a]$ and $\varphi \in [0, 2\pi)$, then in the unloaded state, the punch touches the surface of the elastic half-space at the center of the coordinates, and therefore, the parameter δ_0 can be interpreted as the displacement of the punch under the applied load F. Provided that the punch shape function $\Phi(r, \varphi)$ is known, the equilibrium Equation (4) establishes a relation between the contact force F and the punch displacement δ_0.

Remark 1. *It is pertinent to note here that to have an axisymmetric displacement in the Boussinesq problem, the indenter shape should be axisymmetric. However, an axisymmetric shape of the indenter in unilateral contact does not necessarily imply that the established axisymmetric contact region is circular (see, for example, [17,18]), where indenters of toroidal-type shapes produce an annular contact region, or [19], where the contact region under a non-convex parametric–homogeneous punch is composed from a central circular part and a number of concentric annular regions). On the other hand, the Boussinesq contact problem with a circular area of contact will be non-axisymmetric if the indenter shape function $\Phi(r, \varphi)$ essentially depends on the angular coordinate φ, that is, if the Fourier series (9) contains at least one nontrivial term starting from $n = 1$.*

2.2. Copson's Series Solution

In view of (1), the general solution of Equation (7) can be represented in the form

$$p(r, \varphi) = p_0(r) + \sum_{n=1}^{\infty} p_n^{\mathrm{s}}(r) \sin n\varphi + p_n^{\mathrm{c}}(r) \cos n\varphi. \tag{11}$$

According to Copson (1947), the coefficients of the Fourier series (11) are determined by the following formulas [20]:

$$p_n(r) = -\frac{E^*}{\pi} r^{n-1} \frac{\mathrm{d}}{\mathrm{d}r} \int_r^a \frac{g_n(\rho)\rho \, \mathrm{d}\rho}{\sqrt{\rho^2 - r^2}}, \tag{12}$$

where

$$g_n(\rho) = \frac{1}{\rho^{2n}} \frac{\mathrm{d}}{\mathrm{d}\rho} \int_0^\rho \frac{r^{n+1} f_n(r)}{\sqrt{\rho^2 - r^2}} \, \mathrm{d}r. \tag{13}$$

To be more precise, the coefficients $p_n^{\mathrm{s}}(r)$ and $p_n^{\mathrm{c}}(r)$ are given by Formulas (12) and (13) upon replacing $f_n(r)$ with $f_n^{\mathrm{s}}(r)$ and $f_n^{\mathrm{c}}(r)$, respectively.

We note that in the case $n = 0$, Formulas (12) and (13) represent the Galin–Sneddon solution of the axisymmetric contact problem. The general solution of the non-axisymmetric contact problem in the series from (12) and (13) was also independently derived by Mossakovskii [21].

2.3. Mossakovskii's Series Solution

Under the assumption that the functions $f_0(r)$, $f_n^{\mathrm{s}}(r)$, and $f_n^{\mathrm{c}}(r)$ are continuously differentiable in the interval $(0, a)$, Mossakovskii (1953) simplified Formulas (12) and (13) as follows [21]:

$$p_n(r) = \frac{E^*}{2} \left\{ \frac{C_n r^n}{\sqrt{a^2 - r^2}} - \frac{2r^n}{\pi} \int_r^a \frac{x^{-2n} \, \mathrm{d}x}{\sqrt{x^2 - r^2}} \int_0^x \frac{f_n''(\rho)\rho^{n+1} + f_n'(\rho)\rho^n - n^2 f_n(\rho)\rho^{n-1}}{\sqrt{x^2 - \rho^2}} \, \mathrm{d}\rho \right\}. \tag{14}$$

Here, the constant C_n is given by the formulas

$$C_0 = \frac{2}{\pi} \left(f_0(0) + a \int_0^a \frac{f_0'(\rho) \, \mathrm{d}\rho}{\sqrt{a^2 - \rho^2}} \right), \tag{15}$$

$$C_n = \frac{2}{\pi} a^{1-2n} \int_0^a \frac{f_n'(\rho)\rho^n + n f_n(\rho)\rho^{n-1}}{\sqrt{a^2 - \rho^2}} \, \mathrm{d}\rho, \quad n = 1, 2, \ldots \tag{16}$$

We note that, in light of (3) and (8)–(10), we have $f_0(0) = \delta_0$.

2.4. Leonov's Closed-Form Solution

Let Δ denote the two-dimensional Laplace differential operator, that is

$$\Delta u(r, \varphi) = \frac{\partial^2 u}{\partial r^2} + \frac{1}{r}\frac{\partial u}{\partial r} + \frac{1}{r^2}\frac{\partial^2 u}{\partial \varphi^2}. \tag{17}$$

The general solution to the governing integral Equation (7) in a closed form was first obtained by Leonov (1955). To simplify the writing of his formula, we introduce the notation

$$R = \sqrt{r^2 + \rho^2 - 2r\rho\cos(\varphi - \psi)}. \tag{18}$$

By the definition, R equals the distance between the point of observation (r, φ) and the point of integration (ρ, ψ).

So, according to Leonov, the general solution is given by the following formula [22]:

$$\begin{aligned}
p(r, \varphi) = -\frac{E^*}{2\pi^2} \Bigg\{ &\frac{\pi}{2}\Delta \int_0^{2\pi}\int_0^a \frac{f(\rho, \psi)}{R}\rho\,d\rho\,d\psi \\
&- \int_0^{2\pi}\int_0^a f(\rho, \psi)\left[\frac{1}{R^3}\left(\arctan\frac{\sqrt{a^2 - r^2}\sqrt{a^2 - \rho^2}}{aR} - \frac{\pi}{2}\right) \right.\\
&\left. + \frac{a}{R^2\sqrt{a^2 - \rho^2}\sqrt{a^2 - r^2}}\right]\rho\,d\rho\,d\psi \Bigg\}.
\end{aligned} \tag{19}$$

We note that $\arctan(x) - \pi/2 = -\arctan(1/x)$.

2.5. Mossakovskii's Form of the General Solution

Starting from the series solution (11)–(13), Mossakovskii (1953) derived the following general solution in the following form [23]:

$$\begin{aligned}
p(r, \varphi) = -\frac{E^*}{2\pi^2} \Bigg\{ &\Delta \int_0^{2\pi}\int_0^a f(\rho, \psi)\arctan\left(\frac{\sqrt{a^2 - r^2}\sqrt{a^2 - \rho^2}}{aR}\right)\frac{\rho\,d\rho\,d\psi}{R} \\
&+ \int_0^{2\pi}\int_0^a \frac{af(\rho, \psi)}{\sqrt{(a^2 - r^2)^3(a^2 - \rho^2)}}\frac{(a^4 - \rho^2 r^2)\rho\,d\rho\,d\psi}{\left[a^4 - 2a^2 r\rho\cos(\varphi - \psi) + r^2\rho^2\right]}\Bigg\}.
\end{aligned} \tag{20}$$

Here the same notation is used as introduced by Formulas (17) and (18).

We note that $\arctan(x) - \pi/2 = -\arctan(1/x)$.

2.6. The Galin–Mossakovskii General Solution

We recall that the zeroth term of the Fourier series (9) is defined by the formula

$$f_0(r) = \frac{1}{2\pi}\int_0^{2\pi} f(r, \psi)\,d\psi. \tag{21}$$

As a generalization of the general solution obtained by Galin [24], for the special case, when the contact pressure beneath the indenter is bounded, Mossakovskii (1953) derived the following formula [23]:

$$p(r, \varphi) = \frac{E^*}{\pi} \frac{f_0(0)}{\sqrt{a^2 - r^2}}$$

$$+ \frac{E^*}{2\pi^2} \frac{a}{\sqrt{a^2 - r^2}} \int_0^{2\pi} \int_0^a \frac{(a^4 - \rho^2 r^2)\frac{\partial f}{\partial \rho}(\rho, \psi) + 2a^2 r \sin(\varphi - \psi)\frac{\partial f}{\partial \psi}(\rho, \psi)}{\sqrt{a^2 - \rho^2}\left[a^4 - 2a^2 r\rho \cos(\varphi - \psi) + r^2\rho^2\right]} \, d\rho d\psi$$

$$- \frac{E^*}{2\pi^2} \int_0^{2\pi} \int_0^a \Delta f(\rho, \psi) \arctan\left(\frac{\sqrt{a^2 - r^2}\sqrt{a^2 - \rho^2}}{aR}\right) \frac{\rho \, d\rho d\psi}{R}. \tag{22}$$

We note that, in view of (21), $f_0(r)$ gives the average value of the function $f(r, \varphi)$ on a circumference of radius r. That is why, if $f(r, \varphi)$ is a continuous function, then $f_0(0)$ coincides with the limit of $f(r, \varphi)$ as $r \to 0$.

2.7. Fabrikant's General Solutions

Let us introduce the notation

$$\mathscr{L}(k)g(\varphi) = \frac{1}{2\pi} \int_0^{2\pi} \lambda(k, \varphi - \tau)g(\tau) \, d\tau, \tag{23}$$

where

$$\lambda(k, \tau) = \frac{1 - k^2}{1 + k^2 - 2k \cos \tau}. \tag{24}$$

We note that Formula (23) defines the $\mathscr{L}$-operator [25] that acts on the function $g(\varphi)$, defined on a unit circle.

Moreover, we put

$$\eta = \frac{1}{a}(a^2 - r^2)^{1/2}(a^2 - \rho^2)^{1/2}. \tag{25}$$

According to Fabrikant (1986), the general solution to the governing integral equation of the contact problem under consideration (7), in view of the notation (8), can be represented in the following closed form [25]:

$$p(r, \varphi) = -\frac{E^*}{\pi} \frac{1}{r} \mathscr{L}(r) \frac{d}{dr} \int_r^a \frac{x \, dx}{(x^2 - r^2)^{1/2}} \mathscr{L}\left(\frac{1}{x^2}\right) \frac{d}{dx} \int_0^x \frac{\rho \, d\rho}{(x^2 - \rho^2)^{1/2}} \mathscr{L}(\rho) f(\rho, \varphi). \tag{26}$$

Another form of the Fabrikant solution is given by the following formula:

$$p(r, \varphi) = \frac{E^*}{\pi} \frac{1}{(a^2 - r^2)^{1/2}} \frac{\partial}{\partial a} \int_0^a \frac{\rho \, d\rho}{(a^2 - \rho^2)^{1/2}} \mathscr{L}\left(\frac{r\rho}{a^2}\right) f(\rho, \varphi)$$

$$- \frac{E^*}{2\pi^2} \int_0^{2\pi} \int_0^a \frac{1}{R} \arctan\left(\frac{\eta}{R}\right) \Delta f(\rho, \psi)\rho \, d\rho d\psi. \tag{27}$$

Yet, another form of the Fabrikant solution is given

$$p(r, \varphi) = \frac{E^*}{\pi a} \frac{1}{\sqrt{a^2 - r^2}} \int_0^a \frac{\rho \, d\rho}{\sqrt{a^2 - \rho^2}} \frac{d}{d\rho}\left[\rho\mathscr{L}\left(\frac{r\rho}{a^2}\right) f(\rho, \varphi)\right]$$

$$- \frac{E^*}{2\pi^2} \int_0^{2\pi} d\psi \int_0^a \frac{1}{R} \arctan\left(\frac{\sqrt{a^2 - r^2}\sqrt{a^2 - \rho^2}}{aR}\right) \Delta f(\rho, \psi)\rho \, d\rho. \tag{28}$$

It is pertinent to note here that in terms of the $\mathscr{L}$ operator, the Mossakovskii solution (20) can be represented as follows [26]:

$$p(r, \varphi) = -\frac{E^*}{\pi} \left\{ -\Delta \int_r^a \frac{dx}{(x^2 - r^2)^{1/2}} \int_0^x \frac{\rho \, d\rho}{(x^2 - \rho^2)^{1/2}} \mathscr{L}\left(\frac{r\rho}{x^2}\right) f(\rho, \varphi) \right.$$
$$\left. + \frac{a}{(a^2 - r^2)^{3/2}} \int_0^a \frac{\rho \, d\rho}{(a^2 - \rho^2)^{1/2}} \mathscr{L}\left(\frac{r\rho}{a^2}\right) f(\rho, \psi) \right\}. \tag{29}$$

We also note that, in view of (25), the last terms on the right-hand sides of Formulas (22) and (27) coincide.

3. Contact Pressure at the Center of Circular Contact

For the Fourier series representation (21), it follows that

$$\lim_{r \to 0} p(r, \varphi) = p_0(0). \tag{30}$$

When comparing the series solutions due to Copson (12), (13) and Mossakovskii (14), (15), it is readily seen that only the Mossakovskii series solution allows to evaluate directly the right-hand side of Equation (30).

From Equation (14), it follows that

$$p_0(r) = \frac{E^*}{2} \left\{ \frac{C_0}{\sqrt{a^2 - r^2}} - \frac{2}{\pi} \int_r^a \frac{dx}{\sqrt{x^2 - r^2}} \int_0^x \frac{f_0''(\rho)\rho + f_0'(\rho)}{\sqrt{x^2 - \rho^2}} \, d\rho \right\},$$

so that

$$p_0(0) = \frac{E^* C_0}{2a} - \frac{E^*}{\pi} \int_0^a \frac{dx}{x} \int_0^x \frac{f_0''(\rho)\rho + f_0'(\rho)}{\sqrt{x^2 - \rho^2}} \, d\rho \right\}. \tag{31}$$

By changing the order of integration, we easily transform Formula (31) as follows:

$$p_0(0) = \frac{E^* C_0}{2a} - \frac{E^*}{\pi} \int_0^a \left(\frac{\pi}{2} - \arcsin \frac{\rho}{a} \right) \Delta f_0(\rho) \, d\rho. \tag{32}$$

Here, $\Delta f_0(\rho) = f_0''(\rho) + (1/\rho)f_0'(\rho)$, and C_0 is given by (15). We also note that the integrand in (32) can be further transformed, using the trigonometric formulas

$$\pi/2 - \arcsin x = \arccos x = \arctan(\sqrt{1 - x^2}/x).$$

Finally, we recall that the function $f_0(r)$ is defined by Formula (21).

Now, when comparing the closed-form solutions due to Leonov (19), Mossakovskii (20), and Fabrikant (26) with the Galin–Mossakovskii solution (22) and the Fabrikant solutions (27) and (28), we conclude that only the latter three formulas allow to evaluate directly the contact pressure at the contact center.

By setting $r = 0$ in the Galin–Mossakovskii formula (22), we readily obtain

$$p|_{r=0} = \frac{E^* f_0(0)}{\pi a} + \frac{E^*}{2\pi^2} \int_0^{2\pi} \int_0^a \frac{\partial f}{\partial \rho}(\rho, \psi) \frac{d\rho d\psi}{\sqrt{a^2 - \rho^2}}$$
$$- \frac{E^*}{2\pi^2} \int_0^{2\pi} \int_0^a \Delta f(\rho, \psi) \arctan\left(\frac{\sqrt{a^2 - \rho^2}}{\rho} \right) d\rho d\psi. \tag{33}$$

By taking into account (15) and (21), it can be easily verified that Formulas (32) and (33) are in complete agreement, and they can be rewritten as

$$p\big|_{r=0} = \frac{E^*}{\pi}\left\{\frac{f_0(0)}{a} + \int_0^a \frac{f_0'(\rho)\,d\rho}{\sqrt{a^2-\rho^2}} - \int_0^a \arccos\left(\frac{\rho}{a}\right)\Delta f_0(\rho)\,d\rho\right\},\tag{34}$$

where $f_0(r)$ is defined by Formula (21).

Now, by setting $r=0$ in the Fabrikant solution (27), we obtain

$$p\big|_{r=0} = \frac{E^*}{2\pi^2}\frac{1}{a}\frac{\partial}{\partial a}\int_0^a \frac{\rho\,d\rho}{\sqrt{a^2-\rho^2}}\int_0^{2\pi} f(\rho,\psi)\,d\psi$$

$$-\frac{E^*}{2\pi^2}\int_0^{2\pi}\int_0^a \arctan\left(\frac{\sqrt{a^2-\rho^2}}{\rho}\right)\Delta f(\rho,\psi)\,d\rho d\psi.\tag{35}$$

Here the following formula is used (see Equations (23) and (24)):

$$\mathscr{L}(0)f(\rho,\varphi) = \frac{1}{2\pi}\int_0^{2\pi} f(\rho,\tau)\,d\tau.\tag{36}$$

The first term on the right-hand side of Equation (35) can be simplified as follows:

$$\frac{1}{2\pi}\frac{\partial}{\partial a}\int_0^{2\pi}\int_0^a \frac{f(\rho,\psi)\rho\,d\rho d\psi}{\sqrt{a^2-\rho^2}} = \frac{\partial}{\partial a}\int_0^a \frac{f_0(\rho)\,\rho\,d\rho}{\sqrt{a^2-\rho^2}}$$

$$= \frac{\partial}{\partial a}\left(af_0(0) + \int_0^a f_0'(\rho)\sqrt{a^2-\rho^2}\,d\rho\right)\tag{37}$$

$$= f_0(0) + a\int_0^a \frac{f_0'(\rho)\,d\rho}{\sqrt{a^2-\rho^2}}.$$

Thus, in view of (21) and (38), Formula (35) also completely agrees with Formula (34). Further, from the Fabrikant solution (28), in view of (21) and (36), it follows that

$$p\big|_{r=0} = \frac{E^*}{\pi a^2}\int_0^a \frac{\rho}{\sqrt{a^2-\rho^2}}\left(\frac{d}{d\rho}\left[\rho f_0(\rho)\right]\right)d\rho$$

$$-\frac{E^*}{\pi}\int_0^a \arctan\left(\frac{\sqrt{a^2-\rho^2}}{\rho}\right)\Delta f_0(\rho)\,d\rho,\tag{38}$$

and, taking into account the identity

$$\int_0^a \frac{\rho}{\sqrt{a^2-\rho^2}}\left(\frac{d}{d\rho}\left[\rho f_0(\rho)\right]\right)d\rho = af_0(0) + a^2\int_0^a \frac{f_0'(\rho)\,d\rho}{\sqrt{a^2-\rho^2}},$$

it is readily seen that Formula (38) is equivalent to Formula (34).

4. Contact Stress Intensity Factor

We define the stress-intensity factor (SIF) of the contact stresses as follows:

$$K_{\mathrm{I}}(\varphi) = -\lim_{r\to a}\sqrt{2\pi(a-r)}\,p(r,\varphi).\tag{39}$$

It is to note that the normal stress produced by the punch on the surface points inside the contact area is equal to $-p(r, \varphi)$. It also is worth noting that the contact SIF analysis under a circular punch was considered recently in [16].

4.1. Borodachev's Formula for the Contact SIF

By using the Fabrikant solution (28), Borodachev (1991) derived the following closed-form result [27]:

$$K_{\mathrm{I}}(\varphi) = \frac{E^* a^{1/2}}{2\pi^{3/2}} \int\limits_0^{2\pi} \mathrm{d}\psi \int\limits_0^a \frac{[\rho f(\rho, \psi) - a f(a, \varphi)]\,\mathrm{d}\rho}{\sqrt{a^2 - \rho^2}\,[a^2 - 2a\rho\cos(\varphi - \psi) + \rho^2]}. \tag{40}$$

It is warned that different normalizations can be used in the definition of the SIF.

4.2. Fabrikant's Formula for the Contact SIF

By utilizing his general solution (26) and the general property

$$\lim_{r \to a} \left(\sqrt{a - r}\,\frac{\mathrm{d}}{\mathrm{d}r} \int\limits_r^a \frac{g(x)\,\mathrm{d}x}{\sqrt{x^2 - r^2}} \right) = -\frac{g(a)}{\sqrt{2a}},$$

Fabrikant (1998) derived the following formula [28]:

$$K_{\mathrm{I}}(\varphi) = \frac{E^*}{\sqrt{\pi a}}\,\mathscr{L}\!\left(\frac{1}{a}\right) \frac{\partial}{\partial a} \int\limits_0^a \frac{\rho\,\mathrm{d}\rho}{\sqrt{a^2 - \rho^2}}\,\mathscr{L}(\rho)f(\rho, \varphi), \tag{41}$$

where the $\mathscr{L}$-operator is defined by (23).

4.3. A New Formula for the Contact SIF

Observe that the general solutions (19) and (20) given by Leonov and Mossakovskii contain the application of the Laplace differential operator Δ to the double integral, and therefore, the direct usage of Formula (39) for evaluating the contact SIF is impossible. On the other hand, the Galin–Mossakovskii solution (22) employs the operation of differentiation only under the integral sign.

Let us assume that the function $f(r, \varphi)$ is twice continuously differentiable over the closed circle $0 \le r \le a, 0 \le \varphi < 2\pi$. Then, it can be shown (see Appendix A) that the third term on the right-hand side of Equation (22) is not singular at the contact contour. Hence, from Equations (22) and (39), it follows that

$$K_{\mathrm{I}}(\varphi) = -\frac{E^* f_0(0)}{\sqrt{\pi a}} - \frac{E^* \sqrt{a}}{2\pi^{3/2}} \int\limits_0^{2\pi}\int\limits_0^a \frac{(a^2 - \rho^2)\dfrac{\partial f}{\partial \rho}(\rho, \psi) + 2a\sin(\varphi - \psi)\dfrac{\partial f}{\partial \psi}(\rho, \psi)}{\sqrt{a^2 - \rho^2}\,[a^2 - 2a\rho\cos(\varphi - \psi) + \rho^2]}\,\mathrm{d}\rho\mathrm{d}\psi. \tag{42}$$

Using integration by parts, it can be shown that Formula (40) follows from (42), that is—in other words—Formulas (40) and (42) are equivalent.

5. Discussion

First, we observe that all general solutions considered above make use of both the operations of integration and differentiation. From the computational point of view, it is preferable to avoid differentiating integrals with respect to parameters. That is why, Mossakovskii's series solution (14)–(16) is, in this sense, better than Copson's series solution (12) and (13).

Among the closed-form general solutions, a special interest represents the Galin–Mossakovskii solution (22), since all differentiations are performed under the integral sign. Moreover, in Formula (22), all operations of differentiation are applied directly to the function $f(r, \varphi)$, and thus, are equivalent to differentiating the shape function of the punch $\Phi(r, \varphi)$.

Of note, Fabrikant's solutions (26), (27) and (28) are given by in terms of the $\mathscr{L}$-operator, and their effective application requires the knowledge of its properties (for instance, $\mathscr{L}(k_1)\mathscr{L}(k_2) = \mathscr{L}(k_1 k_2)$).

Another important point to note is the apparent singularity of the Leonov and Mossakovskii general solutions, as each of the terms on the right-hand sides of Equations (19) and (20), generally speaking, has a singularity of the order $(a - r)^{-3/2}$ as $r \to a$. At the same time, the sought-for solution of Equation (7), generally speaking, should possess the square root singularity, that is, the singularity of the order $(a - r)^{-1/2}$ as $r \to a$. This means that the higher-order singularity terms should cancel each other.

It is necessary to note here that the general solutions outlined above hold true in a more general case of a transversely isotropic elastic half-space, provided that the plane of isotropy is parallel to the half-space surface (see, for example, [9]). The potential for further generalization and development of the general solutions presented above relies on the fact that the problem of elastic contact is a core issue in similar contact problems with a circular contact region for functionally graded [29,30], viscoelastic [31,32], thermoelastic [33,34], poroelastic [35,36], magneto-electro-elastic [37,38], multiferroic [39,40] semi-infinite media as well for elastic semi-infinite media with surface effects [41,42].

The main results of the present paper are given by Formulas (34) and (42), which in view of (8), can be rewritten as follows:

$$p\big|_{r=0} = \frac{E^*}{\pi}\left\{ \frac{\delta_0}{a} - \int_0^a \frac{\Phi_0'(\rho)\,\mathrm{d}\rho}{\sqrt{a^2 - \rho^2}} + \int_0^a \arccos\left(\frac{\rho}{a}\right)\Delta\Phi_0(\rho)\,\mathrm{d}\rho \right\}, \tag{43}$$

$$K_{\mathrm{I}}(\varphi) = -\frac{E^*\delta_0}{\sqrt{\pi a}} + \frac{E^*\sqrt{a}}{2\pi^{3/2}}\int_0^{2\pi}\int_0^a \left\{ \sqrt{a^2 - \rho^2}\frac{\partial\Phi}{\partial\rho}(\rho, \psi) \right.$$
$$\left. + \frac{2a\sin(\varphi - \psi)}{\sqrt{a^2 - \rho^2}}\frac{\partial\Phi}{\partial\psi}(\rho, \psi) \right\}\frac{\mathrm{d}\rho\,\mathrm{d}\psi}{\left[a^2 - 2a\rho\cos(\varphi - \psi) + \rho^2\right]}. \tag{44}$$

Here, $\Phi_0'(r)$ is the angle-averaged shape function, i.e.,

$$\Phi_0(r) = \frac{1}{2\pi}\int_0^{2\pi}\Phi(r, \psi)\,\mathrm{d}\psi.$$

It is of interest to note that, in contrast to formulas due to Borodachev (40) and Fabrikant (41), Formula (44) separates the contributions to the contact SIF from the punch displacement δ_0 and the punch shape function $\Phi(r, \varphi)$.

To the best of the author's knowledge, Formulas (43) and (44) have been reported in the literature for the first time.

The Galin–Mossakovskii general solution is recommended for use as the most simple closed-form general solution of the non-axisymmetric Boussinesq contact problem.

Funding: This research received no external funding.

Institutional Review Board Statement: Not applicable.

Informed Consent Statement: Not applicable.

Data Availability Statement: Not applicable.

Acknowledgments: The financial support from the Ba-Yu Scholar program of Chongqing City (China) is gratefully acknowledged. We acknowledge support by the German Research Foundation and the Open Access Publication Fund of TU Berlin.

Conflicts of Interest: The author declares no conflict of interest.

Appendix A. Derivation of the Contact SIF from the Galin–Mossakovskii General Solution

By setting $r = a(1 - \varepsilon)$ in Equation (22), we can rewrite

$$
\begin{aligned}
p\big(a(1 - \varepsilon), \varphi\big) &= \frac{E^*}{\pi a} \frac{f_0(0)}{\sqrt{\varepsilon(2 - \varepsilon)}} \\
&+ \frac{E^*}{2\pi^2} \frac{1}{\sqrt{\varepsilon(2 - \varepsilon)}} \int_0^{2\pi} \int_0^a \frac{\big(a^2 - \rho^2(1 - \varepsilon)^2\big)\frac{\partial f}{\partial \rho}(\rho, \psi) + 2a(1 - \varepsilon)\sin(\varphi - \psi)\frac{\partial f}{\partial \psi}(\rho, \psi)}{\sqrt{a^2 - \rho^2}\big[a^2 - 2a(1 - \varepsilon)\rho\cos(\varphi - \psi) + (1 - \varepsilon)^2\rho^2\big]}\, \mathrm{d}\rho\,\mathrm{d}\psi \\
&- \frac{E^*}{2\pi^2} \int_0^{2\pi} \int_0^a \Delta f(\rho, \psi) \arctan\!\left(\frac{\sqrt{\varepsilon(2 - \varepsilon)}\sqrt{a^2 - \rho^2}}{R_\varepsilon} \right) \frac{\rho\,\mathrm{d}\rho\,\mathrm{d}\psi}{R_\varepsilon},
\end{aligned}
\tag{A1}
$$

where $R_\varepsilon = \sqrt{a^2(1 - \varepsilon)^2 + \rho^2 - 2a(1 - \varepsilon)\rho\cos(\varphi - \psi)}$.

Further, by the definition (39), we have

$$
K_{\mathrm{I}}(\varphi) = -\lim_{\varepsilon \to 0^+} \sqrt{2\pi a \varepsilon}\, p\big(a(1 - \varepsilon), \varphi\big).
\tag{A2}
$$

Hence, by substituting (A1) into Formula (A2) and letting ε tend to zero, we immediately arrive at Formula (42), since the third term on the right-hand side of Equation (A1) is not singular as $\varepsilon \to 0$.

References

1. Sneddon, I.N. The relation between load and penetration in the axisymmetric Boussinesq problem for a punch of arbitrary profile. *Int. J. Eng. Sci.* **1965**, *3*, 47–57. [CrossRef]
2. Borodich, F.M. The Hertz type and adhesive contact problems for depth-sensing indentation. *Adv. Appl. Mech.* **2014**, *47*, 225–366.
3. Barber, J.R. *Contact Mechanics*; Springer: Dordrecht, The Netherlands, 2018.
4. Popova, E.; Popov, V.L. History of "Sneddon" solution in contact mechanics. *PAMM* **2021**, *21*, e202100048. [CrossRef]
5. Popova, E.; Popov, V.L. Ludwig Föppl and Gerhard Schubert: Unknown classics of contact mechanics. *ZAMM-J. Appl. Math. Mech. Angew. Math. Mech.* **2020**, *100*, e202000203. [CrossRef]
6. Schubert, G. Zur Frage der Druckverteilung unter elastisch gelagerten Tragwerken. *Ing.-Arch.* **1942**, *13*, 132–147. (In German) [CrossRef]
7. Leonov, M.Y. On the calculation of elastic foundations. *PMM J. Appl. Math. Mech.* **1939**, *3*, 53–78. (In Russian)
8. Argatov, I.I.; Dmitriev, N.N. *Fundamentals of the Theory of Elastic Discrete Contact*; Polytechnics: St. Petersburg, Russia, 2003. (In Russian)
9. Argatov, I.; Mishuris, G. *Indentation Testing of Biological Materials*; Springer: Cham, Switzerland, 2018.
10. Argatov, I.I.; Sabina, F.J. Contact stiffness indentation tomography: Moduli-perturbation approach. *Int. J. Eng. Sci.* **2020**, *146*, 103175. [CrossRef]
11. Argatov, I.I.; Jin, X.; Keer, L.M. Collective indentation as a novel strategy for mechanical palpation tomography. *J. Mech. Phys. Solids* **2020**, *143*, 104063. [CrossRef]
12. Argatov, I.I. Controlling the adhesive pull-off force via the change of contact geometry. *Philos. Trans. R. Soc. A* **2021**, *379*, 20200392. [CrossRef]
13. Carrier, W.D.; Christian, J.T. Rigid circular plate resting on a non-homogeneous elastic half-space. *Geotechnique* **1973**, *23*, 67–84. [CrossRef]
14. Boswell, L.F.; Scott, C.R. A flexible circular plate on a heterogeneous elastic half-space: Influence coefficients for contact stress and settlement. *Géotechnique* **1975**, *25*, 604–610. [CrossRef]
15. Popov, V.L.; Heß, M.; Willert, E. *Handbook of Contact Mechanics: Exact Solutions of Axisymmetric Contact Problems*; Springer Nature: Berlin, Germany, 2019.
16. Argatov, I. A singularity analysis in Keer's elastic indentation problem. *Mech. Res. Commun.* **2022**, *121*, 103857. [CrossRef]
17. Barber, J.R. Indentation of the semi-infinite elastic solid by a concave rigid punch. *J. Elast.* **1976**, *6*, 149–159. [CrossRef]
18. Argatov, I.; Li, Q.; Pohrt, R.; Popov, V.L. Johnson–Kendall–Roberts adhesive contact for a toroidal indenter. *Proc. R. Soc. A* **2016**, *472*, 20160218. [CrossRef]
19. Borodich, F.M.; Galanov, B.A. Self-similar problems of elastic contact for non-convex punches. *J. Mech. Phys. Solids* **2002**, *50*, 2441–2461. [CrossRef]

20. Copson, E.T. On the problem of the electrified disc. *Proc. Edinb. Math. Soc.* **1947**, *8*, 14–19. [CrossRef]
21. Mossakovskii, V.I. Pressure of a circular punch on an elastic half-space. *Sci. Notes. Inst. Eng. Sci. Autom.* **1953**, *2*, 9–40. (In Russian)
22. Leonov, M.Y. General problem of pressure of a punch of circular shape in planar projection on an elastic half-space. *PMM J. Appl. Math. Mech.* **1953**, *17*, 87–98. (In Russian)
23. Mossakovskii, V.I. General solution of the problem of determining the pressure under the base of a punch circular in plane with no account for friction. *Sci. Notes. Inst. Eng. Sci. Autom.* **1953**, *2*, 41–53. (In Russian)
24. Galin, L.A. Spatial contact problems of the theory of elasticity for punches of circular shape in planar projection. *PMM J. Appl. Math. Mech.* **1946**, *10*, 425–448. (In Russian)
25. Fabrikant, V.I. A new approach to some problems in potential theory. *ZAMM-J. Appl. Math. Mech. Angew. Math. Mech.* **1986**, *66*, 363–368. [CrossRef]
26. Fabrikant, V.I. Complete solutions to some mixed boundary value problems in elasticity. *Adv. Appl. Mech.* **1989**, *27*, 153–223.
27. Borodachev, N.M. Contact problem for an elastic half-space with a near-circular contact area. *Sov. Appl. Mech.* **1991**, *27*, 118–123. [CrossRef]
28. Fabrikant, V.I. Stress intensity factors and displacements in elastic contact and crack problems. *J. Eng. Mech.* **1998**, *124*, 991–999. [CrossRef]
29. Heß, M. A simple method for solving adhesive and non-adhesive axisymmetric contact problems of elastically graded materials. *Int. J. Eng. Sci.* **2016**, *104*, 20–33. [CrossRef]
30. Argatov, I.I.; Sabina, F.J. Recovery of information on the depth-dependent profile of elastic FGMs from indentation experiments. *Int. J. Eng. Sci.* **2022**, *176*, 103659. [CrossRef]
31. Graham, G.A.C. The correspondence principle of linear viscoelasticity theory for mixed boundary value problems involving time-dependent boundary regions. *Q. Appl. Math.* **1968**, *26*, 167–174. [CrossRef]
32. Itou, H.; Kovtunenko, V.A.; Rajagopal, K.R. The Boussinesq flat-punch indentation problem within the context of linearized viscoelasticity. *Int. J. Eng. Sci.* **2020**, *151*, 103272. [CrossRef]
33. Barber, J. Indentation of an elastic half-space by a cooled flat punch. *Q. J. Mech. Appl. Math.* **1982**, *35*, 141–154. [CrossRef]
34. Krenev, L.I.; Aizikovich, S.M.; Tokovyy, Y.V.; Wang, Y.C. Axisymmetric problem on the indentation of a hot circular punch into an arbitrarily nonhomogeneous half-space. *Int. J. Solids Struct.* **2015**, *59*, 18–28. [CrossRef]
35. Yue, Z.; Selvadurai, A. On the asymmetric indentation of a consolidating poroelastic half space. *Appl. Math. Model.* **1994**, *18*, 170–185. [CrossRef]
36. Selvadurai, A.P.S.; Samea, P. On the indentation of a poroelastic halfspace. *Int. J. Eng. Sci.* **2020**, *149*, 103246. [CrossRef]
37. Fabrikant, V.I. Green's functions for the magneto-electro-elastic anisotropic half-space and their applications to contact and crack problems. *Arch. Appl. Mech.* **2017**, *87*, 1859–1869. [CrossRef]
38. Chen, J.; Guo, J. Static response of a layered magneto-electro-elastic half-space structure under circular surface loading. *Acta Mech. Solida Sin.* **2017**, *30*, 145–153. [CrossRef]
39. Chen, W.; Pan, E.; Wang, H.; Zhang, C. Theory of indentation on multiferroic composite materials. *J. Mech. Phys. Solids* **2010**, *58*, 1524–1551. [CrossRef]
40. Wu, F.; Li, X.Y.; Chen, W.Q.; Kang, G.Z.; Müller, R. Indentation on a transversely isotropic half-space of multiferroic composite medium with a circular contact region. *Int. J. Eng. Sci.* **2018**, *123*, 236–289. [CrossRef]
41. Zemlyanova, A.Y.; White, L.M. Axisymmetric frictionless indentation of a rigid stamp into a semi-space with a surface energetic boundary. *Math. Mech. Solids* **2022**, *27*, 334–347. [CrossRef]
42. Argatov, I. The surface tension effect revealed via the indentation scaling index. *Int. J. Eng. Sci.* **2022**, *170*, 103593. [CrossRef]

symmetry

Article

Using Cylindrical and Spherical Symmetries in Numerical Simulations of Quasi-Infinite Mechanical Systems

Alexander E. Filippov [1,2,*] and Valentin L. Popov [2,*]

1   Donetsk Institute for Physics and Engineering, National Academy of Sciences of Ukraine, 83114 Donetsk, Ukraine
2   Technische Universität Berlin, 10623 Berlin, Germany
*   Correspondence: filippov_ae@yahoo.com (A.E.F.); v.popov@tu-berlin.de (V.L.P.)

Abstract: The application of cylindrical and spherical symmetries for numerical studies of many-body problems is presented. It is shown that periodic boundary conditions corresponding to formally cylindrical symmetry allow for reducing the problem of a huge number of interacting particles, minimizing the effect of boundary conditions, and obtaining reasonably correct results from a practical point of view. A physically realizable cylindrical configuration is also studied. The advantages and disadvantages of symmetric realizations are discussed. Finally, spherical symmetry, which naturally realizes a three-dimensional system without boundaries on its two-dimensional surface, is studied. As an example, tectonic dynamics are considered, and interesting patterns resembling real ones are found. It is stressed that perturbations of the axis of planet rotation may be responsible for the formation of such patterns.

Keywords: cylindrical symmetry; spherical symmetry; periodic boundary conditions

Citation: Filippov, A.E.; Popov, V.L. Using Cylindrical and Spherical Symmetries in Numerical Simulations of Quasi-Infinite Mechanical Systems. *Symmetry* **2022**, *14*, 1557. https://doi.org/10.3390/sym14081557

Academic Editor: Sergei D. Odintsov

Received: 15 June 2022
Accepted: 25 July 2022
Published: 28 July 2022

Publisher's Note: MDPI stays neutral with regard to jurisdictional claims in published maps and institutional affiliations.

1. Introduction

Numerical models are often used for solving complex problems in many fields [1–5]. They allow researchers to observe complex systems in artificially created scenarios. This makes it possible to identify the major governing parameters and study the underlying physics. Using these observations, one can later develop sophisticated quantitative models. To be efficient, a heuristic model must account for important features of the phenomenon and its main governing factors.

We often consider it useful to construct mechanical systems from some interacting blocks (or 'particles'). The interactions between them must fulfill certain generic properties, e.g., for solids they should enable the system to sustain its shape, and this should be implemented in the model. The atomic lattice of crystalline solids is responsible for the long-range order in a solid. Its existence is owing to the physics of interatomic interactions. Such a description has general validity and also applies to amorphous solids and even fluids. It also applies to fracture processes, which occur via the subdivision of a solid into fragments before separation. The analogous approach can be applied to describe various phenomena and systems, such as pattern formation in nature, the mixing of powders, and the behavior of dynamic systems (see, for example [6,7]).

At present, it is still not possible to advance to satisfactory length and time scales by using classical molecular dynamics that operate with atoms or molecules. That is why representative objects are often used as particles, whose motion, in its entirety, reflects the pertinent aspects of the behavior of the system. For example, the moveable cellular automata [1], smooth particle hydrodynamics [2], and discrete automata methods [3,6,7] can be mentioned. Note that visualization and processing of the results of the computations are very involved [7]. In this context, the reduction in computation costs is essentially important if one needs a relatively long-term simulation to obtain physically interesting results.

Below we will present one of them, which applies (virtually or naturally) cylindrical and spherical symmetries to reduce the calculation time and memory consumption. Our concern in the present work is primarily the extraction of qualitative information from numerical simulations, with convenient access to direct visualization for various scenarios and over long durations. To do this, we used the isotropic 'moveable automata' technique (cf., e.g., [3,7]), which involves a weak, long-range interaction between particles in addition to a short-range repulsive one.

The form of the interparticle interactions used ensures the existence of a minimum overall potential and the attendant equilibrium state. Such a system does not require 'walls' created by the boundary conditions to form compact fragments of a crystalline lattice. Even in the case when the average density of the material is lower than that required for contiguous space-filling, the particles show a tendency to aggregate [7,8]. With these considerations in mind, we use a model whose details are given below. In solving the equations of motion, we confined ourselves to problems where the thermal energy $k_B T$, where T is the absolute temperature and k_B is the Boltzmann constant, is negligibly small compared to other energy terms involved. However, as is usual in dynamics, mutual transformations between potential and kinetic energy still exist. It can be proven [9] that in the limit of high kinetic energy density, $E_{kin} >> C_{ij}$, where C_{ij} is a stiffness parameter, the system behaves similar to a gas of nearly free-flying particles. In the opposite limit case, $E_{kin} << C_{ij}$, a crystal lattice is formed spontaneously.

The nodes of the lattice oscillate around the bottoms of the potential valleys formed by the neighboring particles. It can be described in terms of a harmonic Hamiltonian, slightly perturbed by non-linear terms. In this sense, the system deforms elastically. While in the gaseous state, the system is isotropic on average, and in the solid state, its lattice has a hexagonal symmetry if the interaction potential between the particles is isotropic. In both limited cases, the system exhibits regular dynamic behavior. However, due to a difference in symmetry, there is a transition between the two states via a mixed disordered state, which, in the presence of dissipation, represents a viscous fluid. Thus, the minimalist model considered accounts for the occurrence of all three aggregate states of the system.

2. Numerical Model

The model describes a system of N particles represented by the vector radius $\mathbf{r}_i$, the momentum $\mathbf{p}_i$, and the interaction potential $U(|\mathbf{r}_i - \mathbf{r}_j|)$ corresponding to the following Hamiltonian:

$$H(\mathbf{r}_i, \mathbf{p}_i) = \sum_{i=1}^{N} \mathbf{p}_i^2 / 2m_i + \sum_{i,j=1}^{N} U(|\mathbf{r}_i - \mathbf{r}_j|)/2. \tag{1}$$

It is opportune to represent the interaction potential by a pair of Gauss potentials:

$$U(|\mathbf{r}_i - \mathbf{r}_j|) = C_{ij} \exp\left\{-[(\mathbf{r}_i - \mathbf{r}_j)/c_{ij}]^2\right\} - D_{ij} \exp\left\{-[(\mathbf{r}_i - \mathbf{r}_j)/d_{ij}]^2\right\}, \tag{2}$$

where C_{ij} and D_{ij} define the magnitude, while c_{ij} and d_{ij} are the radii of attraction and repulsion, respectively. The minimization condition corresponding to an equilibrium reads as follows: $C_{ij} >> D_{ij}$, $c_{ij} < d_{ij}$. Since the parameters C_{ij}, D_{ij}, c_{ij}, and d_{ij} form arrays covering all possible combinations, it can be said, without loss of generality, that the equations of motion

$$m_i \partial \mathbf{v}_i / \partial t = -\partial H(\mathbf{x}_i, \mathbf{p}_i) / \partial \mathbf{p}_i = \mathbf{f}_i, \tag{3}$$

describe systems with any number of components, including one- and two-component systems, which are of interest in this paper. There is a requirement to specify the parameters C_{11}, C_{22}, $C_{12} = C_{21}$, etc.

In standard numerical experiments, the particles are initially placed at random on a rectangle with the side lengths of $[0, L_x]$ and $[0, L_y]$. In addition, in the case of contact with a thermal bath, the boundary conditions need to be accompanied by additional ones. If the temperature T of the thermal bath at a given boundary (or within the system) is non-zero, an array of random δ-correlated Langevin forces need to be included. These are to satisfy the fluctuation-dissipation theorem in the form of

$$\langle \xi(t, x_i, y_i)\xi(t\prime, x_j, y_j)\rangle = D\delta(t - t')\delta_{ij}\langle \xi(t, x_i, y_i)\rangle = 0, \tag{4}$$

where $D = 2\gamma k_B T m/dt$, k_B is the Boltzmann constant, γ is a dissipation constant, and dt is the discrete time step.

Independently of the thermal source, the interaction between the particles produces kinetic energy and causes effective thermalization of the system. So, one has to introduce a dissipation channel acting to equilibrate the relative velocities of particles that happen to be within the distance c_v close to the equilibrium. This can be performed by additive force

$$f_i^{\mathbf{v}} \sim \sum_{j=1}^{N} (\mathbf{v}_i - \mathbf{v}_j)\exp\left[-[(\mathbf{r}_i - \mathbf{r}_j)/c_v]^2\right] \tag{5}$$

acting on the i-th particle, with yet another dissipation constant η. As a result, the equations of motion assume the following form:

$$m_i\frac{\partial \mathbf{v}_i}{\partial t} = f_i^{\mathbf{r}} - \eta f_i^{\mathbf{v}} - \gamma \mathbf{v}_i + \xi(t, \mathbf{r}_i) \tag{6}$$

These equations can be integrated using Verlet's method, which conserves the energy of the system at each time step, provided no energy is supplied externally through mechanical work or temperature variation.

3. Artificial Cylindrical Symmetry

The simplicity of the equations of motion is deceptive. Summation occurs at every time step over all possible neighbors, including very distant ones. At the same time, to obtain a more or less realistic model of the system and exclude the effect of boundary conditions that are too close, one normally needs a sufficiently large number of "particles". It becomes especially important when a long-duration run of the numerical simulation (which often reflects the long duration of the real physical experiment) is needed.

For example, this problem appears when one models a process of strong shear deformation, which causes plastic transformations of the material, which results in so-called soli-state turbulence [9]. If the numerically generated system is too short, movable automata quickly reach a boundary of the specimen in a direction of shear deformation and repulse from it back to the internal regions of the system. It completely modifies the numerically obtained results, which strongly deviate from the picture of the real process. To avoid such an artifact, one has to apply as remote a boundary in the direction of shear as possible. To conserve a reasonably limited number of the automata in the numerical procedure, one has to reduce the width of the system in the orthogonal direction. A system that is too narrow with close boundaries in the orthogonal direction also gives some uncontrollable corrections.

In such a case, it is convenient to employ periodic boundary conditions on the horizontal axis. Accordingly, a particle leaving the interval $[0, L_x]$ is returned to it at the opposite end, and the vectors $\mathbf{r}_i - \mathbf{r}_j$ connect particles located within the interval $[0, L_x]$ or the images of escaped particles at the opposite side of the system. On the vertical axis, the system is limited by plates at $y = 0$ and $y = L_y$ with reflecting boundary conditions: $U_{up} = C\exp[(y - L_y)/c]$ and $U_{down} = C\exp[-y/c]$. The stronger the inequalities $C >> C_{ij}$ and $c << c_{ij}$, the more rigid and sharp the walls are in relation to other forces and lengths relevant to the system.

The conceptual picture of such a system is shown in Figure 1. It is seen directly that two ends of the stripe in the horizontal directions are glued without a break. The particles mostly smoothly pass any (imaginary) gap between two ends of the stripe and continue their motion along a cylindrical surface. In the formal numerical model, the most important thing is that the particles from one side of the stripe interact with the "images" of the particles from another one. However, such a trick is not needed if the particles are placed on the surface of a real cylinder in 3D space.

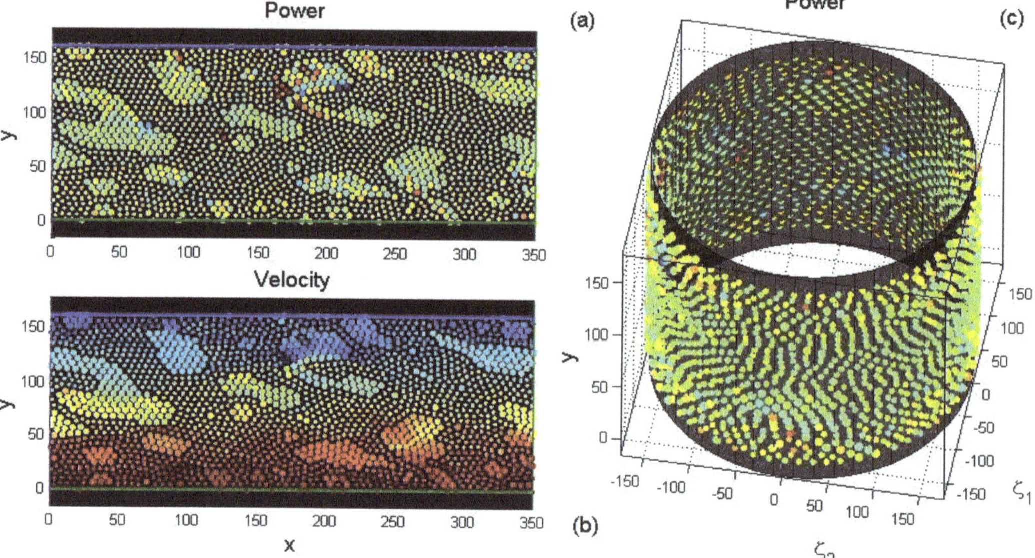

Figure 1. Conceptual picture of a system with periodic boundary conditions in one-dimensional and repulsing boundaries in the orthogonal one. The origin of the particular system and meaning of the colors in different subplots are described in the main text.

In any case, such effective "infiniteness" of the system provides an advantage (sometimes huge) in computer time consumption, because one does not need an extremely long system to minimize the effect of the boundaries, whereas the number of operations grows and squares N^2 of the number of particles N.

Development of the instant structure presented in Figure 1 with the duration is reproduced in the Movie_01.avi [10]. It is seen directly how the particles continue their motion "infinitely" along the cylindrical surface without any disturbance passing any vertical line in subplot (c), including formally connected left and right boundaries of the planar pictures in subplots (a) and (b).

This advantage of the "infiniteness" becomes extremely important when in the extremely long-duration run, one needs to obtain the formation of a nontrivial self-organized system. In particular, mutual collisions of the ordered domains accompanied by their consolidation during the shear of a mixture of two compounds between the plates moving in opposite directions is used in Figure 1 and the corresponding Movie 1.avi [10] as a representative example. This is interesting as a particular realization of solid-state turbulence is discussed in recent publications (see [9] and references therein).

Besides the mathematically formal procedure convenient for the numerical simulations, the described model can correspond to the physically realizable cylindrical system, which potentially gives figurative feedback to the natural experiment, which can reproduce the quasi-infinite run under limited conditions of the laboratory or in some real systems. We will discuss such possibilities below.

4. Practical Example with Stick-Slip Behavior of a System under Shear

For practical needs and a further description, it is useful to calculate a measurable quantity: The external force acting on the system. First, we calculate the work produced by the forces acting on each particle of the system or the corresponding power $p_i = \mathbf{f}_i \mathbf{v}_i$. Summation over all particles yields the total power

$$2VF_{ext}(t) = P(t) = \sum_{i=1}^{N} \mathbf{f}_i \mathbf{v}_i \qquad (7)$$

expended by a pair of external forces to maintain the motion of the boundary plates with a constant velocity, $V = const$. This yields the total force, $F_{ext}(t)$. Calculating spatially distributed power $p_i = \mathbf{f}_i \mathbf{v}_i$ enables one to observe localization of excitations and hence the fragments of the lattice practically move as a whole.

Namely, these colorized spots are clearly seen in some places in Figure 1. They definitely appear in different places and time moments and travel as energy waves along the system following the "events" corresponding to strong changes of the lattice due to mutual collisions of the ordered domains, and so on. It certainly causes avalanches of the energy consumption after stick-slip behavior of the total system under external shear.

Watching the Movie_01.avi, one can observe how (initially randomly placed) relatively small islands of the rigid component collide with one another and form larger ordered "domains", moving and even rotating as a whole. However, they also can split into smaller ones again. The long-duration run allows us to accumulate statistics and observe that the distribution of the domain sizes becomes stationary. Another opposite process is normally organized in the experimental studies [9], where originally, a stripe of the hard material is placed between two soft ones. In this case, it deforms (elastically and plastically) and then splits into a set of islands moving and rotating as a whole. Their distribution tends to be the same equilibrium (stationary, but not static) as the opposite limit. The early infant stage of this process is illustrated in Figure 2, accompanied by the Movie_02.avi [11], reproducing the initial stage of the process in dynamics. One can see the plastic deformation (and destruction into the separate parts later on) of the rigid layer was definitely accompanied by local excitations and even avalanches of them.

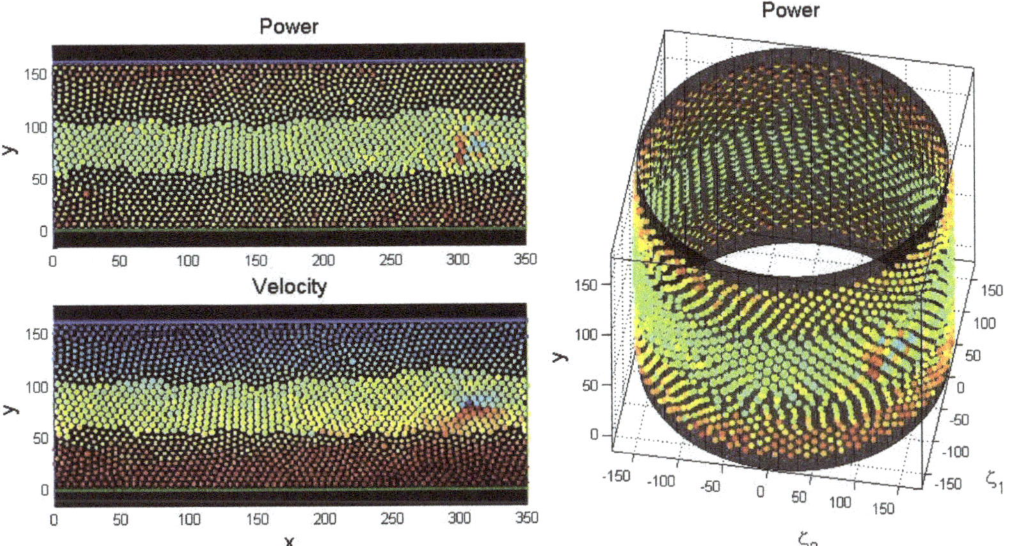

Figure 2. Localized excitation moving along rigid stripe inserted between two soft layers. It is seen directly that velocity of the rigid stripe inserted between two soft ones (green color) is almost constant and much smaller than velocities of the outer layers moving in opposite directions (blue and red colors, respectively).

From academic and practical points of view, it is interesting to learn how this process of a solitary rigid plate (without soft outer ones) behaves itself under shear. This behavior is reproduced in the Movie_3.avi [12].

All the mentioned processes are clearly seen in dynamics in the sequence of the movies Movie_03.avi, Movie_04.avi, and Movie_05.avi [12–14] where direct observation is accompanied by the curves giving quantitative information during the long (potentially infinite) run. In particular, one can see how the time–space map of the horizontal velocity of a selected vertical cross-section recording is static. It records information about the history of the events (wave propagation and their avalanches) in the system. For a two-component system, one can numerically analyze images from every frame of the movie and calculate the time depending on the number of domains in the rigid phase, the histogram of the area occupied by the domains of a given size, as well as the size of the maximal and averaged domain shown to accumulate during the long-duration run. Everything is reproduced directly in the movies. Analogous information can be accumulated in the static form and reproduced, as the examples in Figures 3–5 show below.

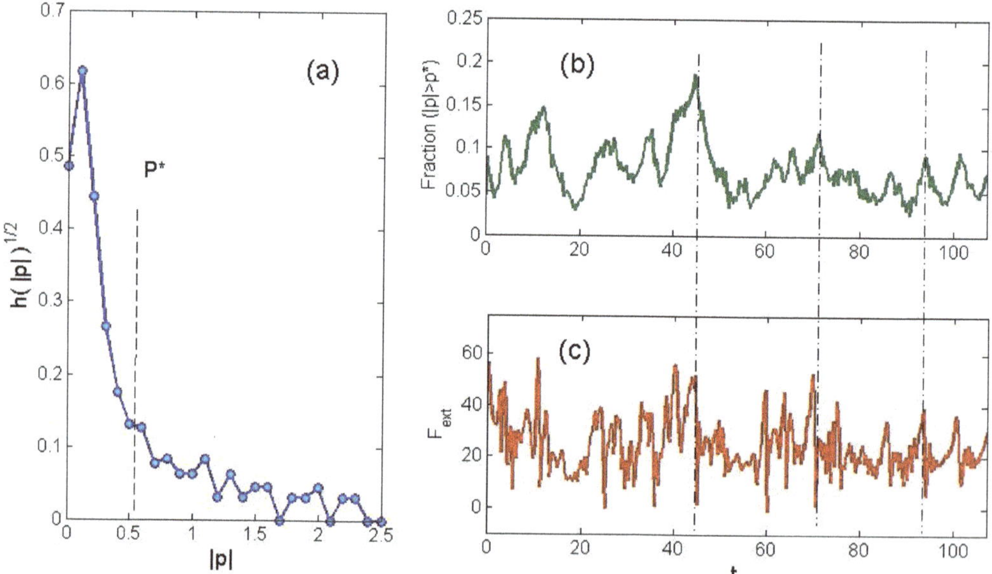

Figure 3. Square root $h\left(|p_i|\right)^{1/2}$ of the histogram $h(|p_i|)$ (**a**) and the time dependence of the fraction of the points for which the inequality $|p_i| > p*$ holds (**b**). Note the bursts in the magnitude of this fraction corresponding to the peaks in the time dependence of the total force $F_{ext}(t) = P(t)/V$, which is presented in (**c**).

Let us specifically mention once more that sliding and jumping over a number of contacting "lattice planes", which have different orientations, results in a phenomenon that can be treated as a kind of "stick-slip" behavior in the system under consideration.

Besides direct visualization, it is useful to characterize the excited sites by the absolute value of the power, $|p_i| = |\mathbf{f}_i \mathbf{v}_i|$. For that, we plot a histogram $h(|p_i|)$ for the probability of finding a given value of $|p_i|$. As expected, the function $h(|p_i|)$ has a pronounced maximum at low values of $|p_i|$ associated with small perturbations of the system almost everywhere within the ordered domains. The regions of large strain contribute to the total power

$$2VF_{ext}(t) = P(t) = \sum_{i=1}^{N} \mathbf{f}_i \mathbf{v}_i$$

most significantly, but the partial contribution of an individual region is not very significant. As a result, the distribution $h(|p_i|)$ has an extended 'tail' at large $|p_i|$.

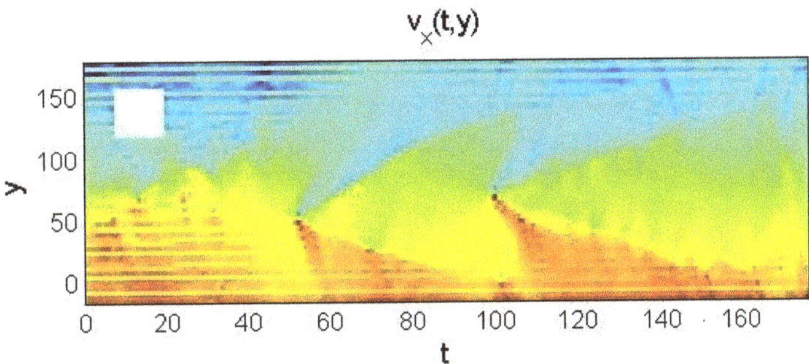

Figure 4. Time–space map of the horizontal velocity of selected vertical cross-section $v_x(t, y)$ recording static information about history of the events (wave propagation) in the system.

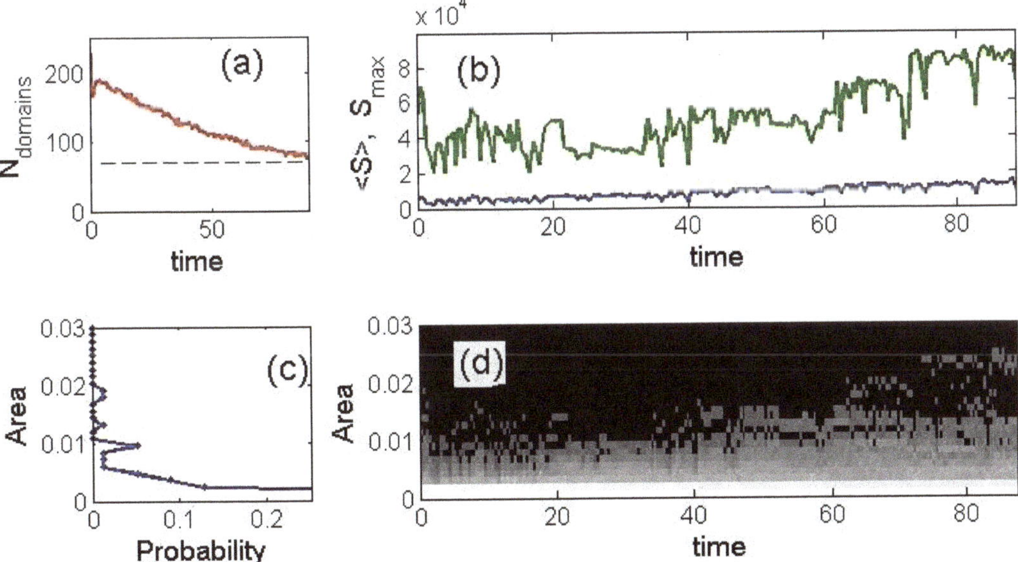

Figure 5. Time-dependent number of rigid phase domains, histogram of area occupied by the domains of the given size, shown accumulated during long-duration run, as well size of the maximal and averaged domain.

For both the locations of the maxima and the asymptotes to be well-resolved at the same time, a nonlinear function $h\left(|p_i|\right)^{1/2}$ is plotted in Figure 3. It can be roughly divided into two regions—left and right of a vertical dashed line that marks a certain characteristic value $p*$. The fraction of points where $|p_i| > p*$ holds, i.e., those points where intensive processes occur, evolves with time. The results of this procedure for a particular realization of the deformation process are shown in Figure 3b. One can recognize pronounced bursts of activity, which can be associated with the periods of emergence and development of rebuilt structures localized at the edges of domains (microscopic analogues of the earthquakes) seen in Movie_3.avi [12]. The same periods can also be identified in the time dependence of the total picture presented in Figure 3c. To facilitate the comparison, some of the bursts are marked with dash-dotted lines.

The function $F_{ext}(t)$ represents the time dependence of the force required to slide a solid on a surface. From this dependence, the yield force, or yield stress, can be identified.

In our computer experiments, this dependence resembled that of the resistance force for sliding a solid on a surface with random roughness in the stick-slip regime.

5. Realistic Quasi-Infinite Objects

Now let us discuss some physical objects with cylindrical and spherical symmetries, which can serve as examples of effectively infinite systems. First of all, quite often, long-run shear-force deformations are organized by means of rotating cylindrical systems. The typical configuration of such a system is illustrated in Figure 6 below. Here, an experimental material is placed between two cylindrical boundaries. The central one is motionless and the external one is rotating. It causes a configuration where different layers of the material inside tend to rotate with the velocity, which gradually increases from $V(R_{\min}) \simeq 0$ to $V(R_{\max}) \simeq V_{ext}$. If both radii are large enough (and in following the system of the coordinates), it is mechanically equivalent to the motion between two plates moving with the velocities $V_{\max} = -V_{ext}/2$ and $V_{\max} = V_{ext}/2$. Another interesting example of a naturally cylindrical system can also be seen in biology [9].

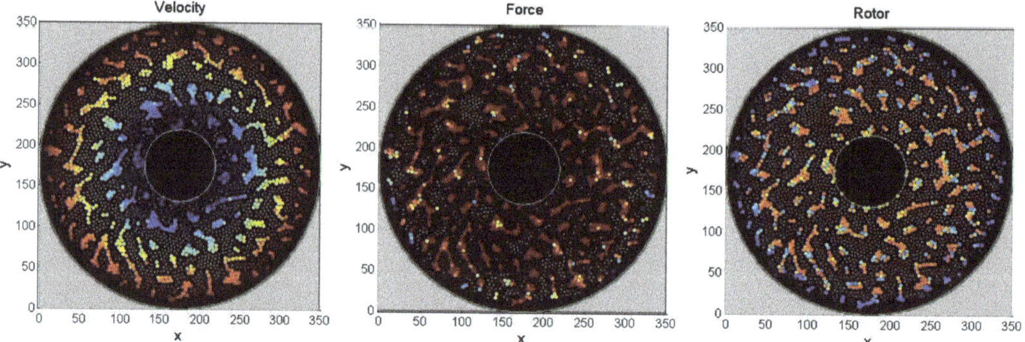

Figure 6. Instant configuration of the cylindrical quasi-infinite system presented in the Movie_6.avi [15].

This fact is clearly seen in the left subplot of Figure 6 where the particles are colored depending on their velocity (from maximum red to minimum blue, respectively). Movie_06.avi [15] reproduces the process of formation and growth of the rigid domains inside the two-component mixture and the tendency towards the dynamic equilibrium distribution of their sizes. The process strongly resembles the same process in the abstract cylindrical system studied above. It also can be continued, as far as necessary.

The only (but very important) difference that such a system can show exists in reality. It even has its own advantage that the numerical procedure in such a configuration does not need fictional images. So, it is potentially expected to require less computer memory and be less time-consuming. However, it still has the disadvantage that due to the geometry, the inner and outer layers of the system are not exactly equivalent. To obtain better equivalence, one has to use both relatively large external and internal radiuses. It actually demands a relatively large number of mutually interacting particles (movable automata).

Spikes of the interaction forces in the places where the domains collide or strongly deform are seen directly in the middle subplot of Figure 6. There is one more interesting characteristic of the system that has to be elucidated. The system definitely rotates as a whole. At the same time, different regions mutually interact and can rotate quasi-independently. This effect is practically the same as that which can be observed in the abstract quasi-infinite system above.

To separate integral and differential rotations, it is convenient to directly calculate the local circulation (below, for briefness, we call it 'rotor' and denote it as rot_j) of the particles in the proximity of every particle of the system and colorize them accordingly. According to this local circulation, the particle in position $\mathbf{r}_j$ is given by the following equation:

$$rot_j = \sum_{k}^{N_{proxy}} [(\mathbf{v}_k - \mathbf{v}_j) \times (\mathbf{r}_k - \mathbf{r}_j)]/(|\mathbf{r}_k - \mathbf{r}_j|). \tag{8}$$

Differential rotation is visualized in the right subplot of Figure 6, and is especially clearly seen in the dynamics of the corresponding Movie_06 [15]. One can distinguish plenty of places where the particles surrounding a given particle instantly rotate in the direction opposite to the rotation of the complete system. This effect corresponds to the "solid-state turbulence", which was already observed in a system confined between two planar plates moving in alternative directions.

The mentioned rotation is even more pronounced if the space between the internal and external cylinders is not so densely filled by the soft component. In this case, the "rocks" formed by the hard component aggregate faster and have more empty space to rotate as a whole. An example of such a system is reproduced in Movie_07.avi [16]. Here, one can clearly observe the rotation of the solid "rocks" accompanied by the motion of their sides colored by blue in the left subplot, which marks the particles playing a role in the instant centers around which their neighbors locally rotate in the opposite direction. Besides, the collisions between the rocks are clearly marked by the flashes of force colors shown in the central subplot.

Let us note in advance that the visualization by means of the colorized rotor is also very useful for the study of rotating systems with spherical symmetry. In the context of the present paper, such symmetry is another realization of a quasi-infinite system existing in nature.

It appears when a spherical belt near the surface of the sphere can rotate more or less independently of the rotation of the whole object, for example, when it is an atmosphere or lithosphere of a planet. In both cases, the belts on the surface are partially involved in the rotation either by an effect (in some senses, just physical friction) on the solid surface of the planet or by the continental flotation on magma.

It is important to mention that it causes practically the same configuration of the velocities, which was studied above for the shear-induced turbulence. In reality, the velocity of the spherical belt monotonously depends on the latitude. It is maximal near the equator of the planet and has to decrease down to zero near the poles. In fact, one has a doubled picture of the shear-induced motion with two stripes mirroring one another.

Despite the complication related to the principally 3D configuration, it paradoxically gives an advantage from the computing point of view. It does not need phantom images of the particles, because the boundary does not exist at all. Certainly, it allows the simulation to run infinitely without any limitation from the boundary conditions. So, one can obtain rather realistic results using a relatively small number of movable automata.

The only thing one needs is to add an attraction (gravitation in fact) of the particles to the solid spherical surface. In principle, this force is a combination of the Newtonian gravitation potential $U_{attract} \sim 1/|r_j - R|$ and exponential repulsion form the surface $U_{repuls} \sim \exp[(r_j - R)/c]$. However, in the particular model, one needs only to "glue" the particles to the surface, so the cheapest way is to apply sufficiently strong confinement with the potential $U_{conf} \sim (r_j - R)^2$.

Generally speaking, in the case of a spherical surface, the definition $rot_j = \sum_{k}^{N_{proxy}} [(\mathbf{v}_k - \mathbf{v}_j) \times (\mathbf{r}_k - \mathbf{r}_j)]/(|\mathbf{r}_k - \mathbf{r}_j|)$ produces a vector in 3D space. To associate it with the vortices and anti-vortices in the spherical belt (of atmo- or lithosphere) we visualize its projection perpendicular to the surface with the coordinate $\mathbf{r}_j$ and colorize it using the standard MatLab map. It is expected that, due to the general rotation of the 'planet', in the case of ideal laminar flow, the picture will have generally blue and red belts of color on the southern and northern hemispheres and a yellow–green stripe around the equator.

However, if the vortices do exist, they will appear as spots of alternative colors (red in southern and blue in northern hemispheres, respectively).

For a trial, we start from the surface randomly covered by the relatively soft material (viscous "atmosphere") and allow it to self-organize due to the rotation. As expected, we obtain something similar to layered rotation with some vortices appearing between the layers. This picture reminds us of a Jupiter-like structure. Figure 7 reproduces a sequence of early configurations of the self-organization in a spherical quasi-infinite system. The first reproduces the very beginning of the process, starting from the randomly distributed particles on the sphere (a), followed by two intermediate ones (b)–(c), and the last subplot (d) reproduces a picture close to the stationary moving configuration. In this stage, the parallel belts of the vortices and anti-vortices in the "equatorial" region (shown by red and blue colors, respectively) are already seen.

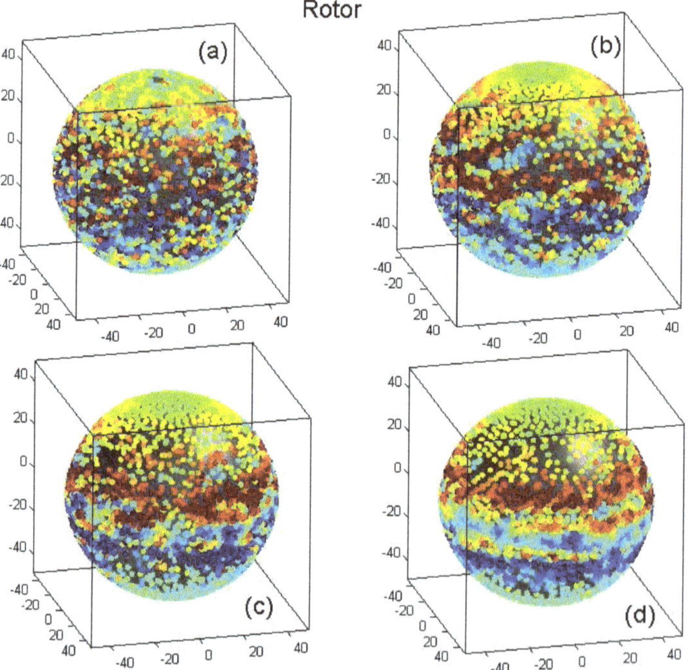

Figure 7. Sequence of the instant configurations of self-organized spherical quasi-infinite system. Early (starting from the randomly distributed particles on the sphere), two intermediate, and close to the stationary moving configurations are shown in the consequent subplots (**a–d**), respectively. Parallel belts of the vortices and anti-vortices in the "equatorial" region corresponding to the positive and negative rotor projection perpendicular to the spherical surface (shown by red and blue colors, respectively) are seen directly.

Depending on the relations between the interaction and damping coefficients, as well as the speed of the rotation, one can obtain a different stationary picture. The vantage of the model is that, in principle, one can run it for as long as necessary to obtain rich stationary behavior. We performed a number of such experiments and obtained sometimes very realistic patterns, which closely resembled recently observed images of planetary atmospheres. See, for example, Movie_08.avi [17].

It is important to note that (as it already was observed in the 2D case), depending on the values of interaction amplitudes C_{ij}, D_{ij} and distances c_{ij}, d_{ij} the equations of motion can describe either soft (liquid) materials, solid ones, or even a mixture of solid and liquid ones. In the context of a spherical system, it provides an intriguing opportunity to model a flotation of the "continental plates" involved in the motion of a rotating liquid "magma".

Certainly, we understand that the model cannot cover all the aspects of the extremely long and multifactorial planetary formation and tectonics. However, in the context of the present conceptual work, one can claim that the model has its own advantages. It is simple, very efficient, and allows for studying a large number of the "theoretical toys" with different relations between all the constants, randomly chosen initial conditions, and scenarios, including very exotic ones. Despite the wide variety of these scenarios, we already observed more or less typical invariants.

One of the typical intermediate stages created by the process reproduced in Movie_09.avi [18] is shown in Figure 8, where the spatial distribution of the rigid and soft (brown and blue, respectively) components is reproduced in subplot (a). Corresponding to the present configuration, distributions of the local velocities and forces are shown in subplots (b) and (c), respectively. In particular, deep red spots of the local forces demonstrate some kind of "earthquakes", which normally appear when the "continents" either collide one with another or the "continents" are torn apart.

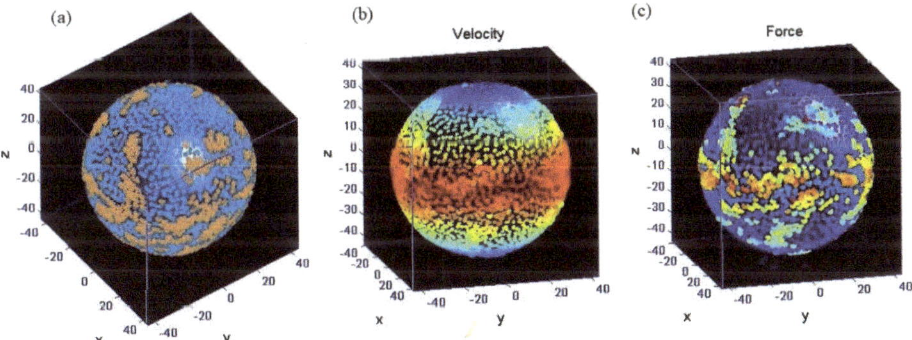

Figure 8. Typical intermediate configuration of the two-component spherical system reproduced in Movie_09. Spatial distribution of the rigid and soft (brown and blue, respectively) components is reproduced in subplot (**a**). Corresponding to the present configuration, distributions of the local velocities and forces are shown in subplots (**b,c**), respectively. A tendency of the concentration of the "continents" to "equator" at fixed axis of the rotation is seen directly.

Our observations of the different scenarios testify to the general tendency of the concentration of the "continents" to the "equator" at a fixed axis of rotation. As a rule, after the long run, they finish in either a kind of "Pangea", or perhaps in two to three "continents" of very different sizes. It partially resembles the real geological history of the Earth, but partially differs from it, because it is known that "Pangea", over time, broke into several smaller parts, which drifted from the equatorial region.

One of the possible reasons for such catastrophes can be provoked by a periodic change in the direction of rotation of the planet. It can be caused by several celestial reasons, due to interplanetary perturbations or even dramatic collisions between planets. One can mention as an example Uranium, which, in contrast to the majority of planets, currently rotates around an axis lying almost in the plane of ecliptics.

We simulated such dramatic scenarios using periodic changes in the direction of the rotational axis. One of them is reproduced in Movie_10.avi [19]. It shows a sequence of "geological" catastrophes, which are accompanied by huge surges of "geological activity". These catastrophes are separated by relatively long calm periods, during which the system normally relaxes to a small number of "continents", which form sometimes extremely realistic global patterns. Some of them strongly resemble the current pattern of our planet Earth, while others reproduce the pictures known from recorded geological history.

6. Conclusions

We presented conceptual models illustrating how cylindrical and spherical symmetries can be applied to reduce the time consumption of numerical simulations in the many-body problem. Often, to obtain practically and theoretically interesting results regarding many-body systems, one needs an extremely long run of corresponding numerical procedures. However, in a normal case, sufficient space for such a run supposes a rather extended system with plenty of interacting particles. It is shown in the present study that periodic boundary conditions corresponding to formal cylindrical symmetry allow for avoiding the problem of a huge number of interacting particles. This trick minimizes the effect of limited boundary conditions and still allows one to obtain reasonably correct results that are interesting from a scientific point of view.

In the second part of the paper, we presented a physically realizable cylindrical configuration and analyzed its advantages and disadvantages for both numerical and physical aspects. Furthermore, the spherical symmetry was studied. In particular, it is stressed that the 2D surface (belt) of the 3D spherical body is a unique natural example of a system, without boundaries at all.

Finally, we present some interesting patterns, which resemble those known from the tectonic history of our planet. It is shown that very realistic continental patterns can be obtained under supposition about perturbations of the planet rotation due to external interplanetary interactions or other extraterrestrial reasons.

Author Contributions: Investigation, A.E.F.; Methodology, V.L.P.; Visualization, A.E.F.; Writing—original draft, A.E.F.; Writing—review & editing, V.L.P. All authors have read and agreed to the published version of the manuscript.

Funding: This research received no external funding.

Institutional Review Board Statement: Not applicable.

Informed Consent Statement: Not applicable.

Data Availability Statement: Not applicable.

Acknowledgments: We acknowledge support by the German Research Foundation and the Open Access Publication Fund of TU Berlin.

Conflicts of Interest: The authors declare no conflict of interest.

References

1. Hoover, W.G. *Smooth Particle Applied Dynamics. The State of the Art*; Advanced Series in Nonlinear Dynamics; World Scientific: Singapore, 2006; Volume 25.
2. Monaghan, J. Smoothed particle hydrodynamics and its diverse. *Annu. Rev. Fluid Mech.* **2012**, *44*, 323–346. [CrossRef]
3. Dmitriev, A.I.; Nikonov, A.Y.; Filippov, A.E.; Psakhie, S.G. Molecular dynamics study of the evolution of rotational atomic displacements in a crystal subjected to shear deformation. *Phys. Mesomech.* **2019**, *22*, 375–381. [CrossRef]
4. Persson, B.N.J. *Sliding Friction, Physical Properties and Applications*; Springer: Berlin/Heidelberg, Germany, 2000.
5. Popov, V.L. *Contact Mechanics and Friction Physics*; Springer: Berlin/Heidelberg, Germany, 2010; p. 362.
6. Filippov, A.E.; Popov, V.L. Numerical simulation of dynamics of block media by movable lattice and movable automata methods. *Facta Universitatis. Series: Mechanical Engineering.* 2022. Available online: http://casopisi.junis.ni.ac.rs/index.php/FUMechEng/article/view/10626 (accessed on 20 July 2022).
7. Filippov, A.E.; Gorb, S.N. *Combined Discrete and Continual Approaches in Biological Modelling*; Springer: Berlin/Heidelberg, Germany, 2020.
8. Filippov, A.E.; Guillermo-Ferreira, R.; Gorb, S.N. "Cylindrical worlds" in biology: Does the aggregation strategy give a selective advantage? *Biosystems* **2019**, *175*, 39–46. [CrossRef] [PubMed]
9. Beygelzimer, Y.; Filippov, A.E.; Kulagin, R.; Estrin, Y. Turbulent shear flow of solids. *arXiv* **2021**, arXiv:2111.05148.
10. Filippov, A.É.; Popov, V.L. Using Cylindrical and Spherical Symmetries in Simulations—Video 1. Available online: https://doi.org/10.13140/RG.2.2.13988.35207 (accessed on 20 July 2022).
11. Filippov, A.É.; Popov, V.L. Using Cylindrical and Spherical Symmetries in Simulations—Video 2. Available online: https://doi.org/10.13140/RG.2.2.20699.23843 (accessed on 20 July 2022).
12. Filippov, A.É.; Popov, V.L. Using Cylindrical and Spherical Symmetries in Simulations—Video 3. Available online: https://doi.org/10.13140/RG.2.2.13464.06408 (accessed on 20 July 2022).

13. Filippov, A.É.; Popov, V.L. Using Cylindrical and Spherical Symmetries in Simulations—Video 4. Available online: https://doi.org/10.13140/RG.2.2.26885.83682 (accessed on 20 July 2022).
14. Filippov, A.É.; Popov, V.L. Using Cylindrical and Spherical Symmetries in Simulations—Video 5. Available online: https://doi.org/10.13140/RG.2.2.20174.95044 (accessed on 20 July 2022).
15. Filippov, A.É.; Popov, V.L. Using Cylindrical and Spherical Symmetries in Simulations—Video 6. Available online: https://doi.org/10.13140/RG.2.2.25208.11520 (accessed on 20 July 2022).
16. Filippov, A.É.; Popov, V.L. Using Cylindrical and Spherical Symmetries in Simulations—Video 7. Available online: https://doi.org/10.13140/RG.2.2.18497.22887 (accessed on 20 July 2022).
17. Filippov, A.É.; Popov, V.L. Using Cylindrical and Spherical Symmetries in Simulations—Video 8. Available online: https://doi.org/10.13140/RG.2.2.21852.67203 (accessed on 20 July 2022).
18. Filippov, A.É.; Popov, V.L. Using Cylindrical and Spherical Symmetries in Simulations—Video 9. Available online: https://doi.org/10.13140/RG.2.2.35274.44485 (accessed on 20 July 2022).
19. Filippov, A.É.; Popov, V.L. Using Cylindrical and Spherical Symmetries in Simulations—Video 10. Available online: https://doi.org/10.13140/RG.2.2.10947.48166 (accessed on 20 July 2022).

MDPI

Correction

Correction: Popov, V.L. An Approximate Solution for the Contact Problem of Profiles Slightly Deviating from Axial Symmetry. *Symmetry* 2022, 14, 390

Valentin L. Popov [1,2]

1   Department of System Dynamics and Friction Physics, Technische Universität Berlin, 10623 Berlin, Germany; v.popov@tu-berlin.de
2   National Research Tomsk State University, 634050 Tomsk, Russia

Citation: Popov, V.L. Correction: Popov, V.L. An Approximate Solution for the Contact Problem of Profiles Slightly Deviating from Axial Symmetry. *Symmetry* 2022, 14, 390. *Symmetry* **2022**, 14, 2108. https://doi.org/10.3390/sym14102108

Received: 19 August 2022
Accepted: 30 September 2022
Published: 11 October 2022

Publisher's Note: MDPI stays neutral with regard to jurisdictional claims in published maps and institutional affiliations.

Text Correction

There were misprints in Equations (40), (65), (66), and (67) in the original publication [1]. The correct Equation (40) of the original publication is:

$$p(r,\varphi) = \frac{E^*}{\pi} \int_r^{a(\varphi)} \frac{\widetilde{a}(\varphi)}{\sqrt{\widetilde{a}(\varphi)^2 - r^2}} \frac{1}{\widetilde{a}_0} \frac{\mathrm{d}g_0(\widetilde{a}_0)}{\mathrm{d}\widetilde{a}(\varphi)} \mathrm{d}\widetilde{a}(\varphi) = \frac{2}{\pi} E^* (2d \cdot \overline{\psi})^{1/2} \sqrt{1 - \left(\frac{r}{a(\varphi)}\right)^2} \quad (40)$$

The correct form of Equations (65) of the original publication is:

$$\gamma(a) = a \int_0^a \frac{n r^{n-1}}{\sqrt{a^2 - r^2}} \mathrm{d}r = \kappa_n a^n, \ \kappa_n = \int_0^1 \frac{\zeta^{n-1} \mathrm{d}\zeta}{\sqrt{1 - \zeta^2}} = \frac{\sqrt{\pi}}{2} \frac{n\Gamma\left(\frac{n}{2}\right)}{\Gamma\left(\frac{n}{2} + \frac{1}{2}\right)} \quad (65)$$

The correct form of Equations (66) of the original publication is:

$$\delta g_\varphi(a) = \kappa_n a^n \left(\psi(\varphi) - \overline{\psi}\right), \ \delta G_\varphi(a) = \kappa_n \frac{a^{n+1}}{n+1} \left(\psi(\varphi) - \overline{\psi}\right) \quad (66)$$

The correct form of Equation (67) of the original publication is:

$$a(\varphi) = a_0 \left(1 + \frac{n+2}{n(n+1)} \left(1 - \frac{\psi(\varphi)}{\overline{\psi}}\right)\right) \quad (67)$$

The author apologizes for any inconvenience caused and state that the scientific conclusions are unaffected. This correction was approved by the Academic Editor. The original publication has also been updated.

Reference

1.   Popov, V.L. An Approximate Solution for the Contact Problem of Profiles Slightly Deviating from Axial Symmetry. *Symmetry* **2022**, 14, 390. [CrossRef]

MDPI

St. Alban-Anlage 66

4052 Basel

Switzerland

Tel. +41 61 683 77 34

Fax +41 61 302 89 18

www.mdpi.com

Symmetry Editorial Office

E-mail: symmetry@mdpi.com

www.mdpi.com/journal/symmetry